C0 1 72 11637 BB

P
Fi
Th

Praise for previous editions of Dinosaurs and the Expanding Earth ...

Stephen Hurrell's *Dinosaurs and the Expanding Earth* is something completely different. Hurrell's thesis - yes, this is an original work rather than a re-presentation of existing knowledge - is that large dinosaurs were forced to give way to smaller mammals because the Earth's gravity has increased, making the life of large creatures untenable.

... his thesis is well presented. ...

Geology Today Review

(This) cleared up a very contentious issue for me ... I checked my mathematical modeling today and discovered ... surface gravity during the Permian was about 50% what it is today, precisely what you are suggesting.

Dr James Maxlow, Geologist and author of Terra Non Firma Earth

Have you seen the huge dinosaur's footprint in Colorado? Or in Australia? Have you looked critically at a dinosaur skeleton in a Paleontology Museum (If you visit the Cappellini Museum in Bologna)? The size, posture and estimated weight of these giants are impressive and the problem of the mechanical deambulation of such large bodies has been posed many times. ...

The book is written in a plain straightforward style ...

Its clear and lively descriptions lead the reader straight to the core of the arguments. ...

Review in ANNALS OF GEOPHYSICS by Dr Giancarlo Scalera.

For more reviews go to: www.oneoffpublishing.com

Dinosaurs and the Expanding Earth

Dinosaurs and the Expanding Earth

Stephen Hurrell

All rights reserved. No part of this publication may be reproduced by any means, or transmitted, or translated without the written permission of the publisher, excepting brief quotes used for reviews.

**Hardback Third Edition
ISBN:10 0952 26037 9
ISBN:13 9780952 26037 0**

Third Edition published 2011

Copyright © 1994 Stephen William Hurrell
Additional material © 2003 & 2011 Stephen William Hurrell

Illustrations copyright © 1994 Stephen William Hurrell
Additional material © 2003 & 2011 Stephen William Hurrell
except for some dinosaur illustrations copyright Selectafont
and where indicated otherwise.

The rights of the author have been asserted.

British Library Cataloguing-in-Publication Data.
A catalogue record for this book is available from the British Library.

First Edition Published in 1994
ISBN:10 0952 260301
Reprinted 1996 and 2001
Second Edition (ebook edition) Published in 2003
ISBN:10 0952 26031X

Published by Oneoff Publishing.com
www.oneoffpublishing.com

Acknowledgements

I would like to thank everyone who has contributed to the concepts expressed in this book. I have now received hundreds of comments and suggestions about a reduced gravity on an ancient Earth which I initially presented in the first editions of this book, and these have helped make this third edition a better book. I would particularly like to thank John Davidson, James Maxlow, Ramin Amir Mardfar, Bill Erickson, Giancarlo Scalera, Neal Adams and Larry Myers for spending their valuable time providing helpful comments on this third edition. Sam Warren Carey and Lance Endersbee also commented on the first edition and although they have since passed away I am still grateful for their advice. Rod Townend and Robert Tipping both read drafts of the first edition and their efforts are still noticeable and appreciated for this third edition. Many thanks must go to Helge Hilgenberg for giving me permission to use a photograph of her father's expanding earth globe that was first published in his 1933 book. Also I am very grateful to Werner Kraus for the wonderful photographs he has provided of his life-sized reconstructions of a giant dragonfly and millipede from the Carboniferous. They very clearly show the large scale of life in those times. Any mistakes in this book remain mine as are the explanations and opinions given.

As in the first two editions of this book, I have tried to write it as clearly as possible even though it covers a very broad range of scientific concepts. It would be pointless making this book too technical since an expert in one subject is a layman in others and the book must cover a wide range of scientific disciplines. I have tried to provide a note in the text when concepts and illustrations have been inspired by previous authors so readers who wish to pursue a particular subject in greater detail can refer to the original publications mentioned. Special thanks go to my wife for her continuing role as proof reader and critic.

The Author

Stephen Hurrell lives near Liverpool in the UK. He has worked in different mechanical engineering design positions for various companies. It was his role as a mechanical design engineer at the UK's Electricity Research Centre that first offered him his insight into how scale effects were pertinent to the biomechanical problems of the dinosaurs' large size. These thoughts about dinosaurs as engineering structures, and the problems of scale effects, fostered the development of the Reduced Gravity Earth theory and its implications for the Expanding Earth. He can be contacted through his web site www.dinox.org.

Contents

Acknowledgements	7
The Author	8
Contents	9
Figures	13
Introduction	19
1 - The Scale Limits of Life	29
The Paradox of Scale and Lifestyle	31
Long Necks	33
Previous Attempts to Solve the Mystery	34
The Size Limits on Today's Life	36
The Effects of Scale on all Structures	36
Life's Changing Form due to Scale Effects	39
The Optimum Form	41
Optimum Form for Different Scales	42
Flight	45
Plants	47
A Solution to the Giants of the Past	48
The Relationship between Scale and Gravity	50
2 - The Giants of the Past	53
Giant Plants of the Palaeozoic Era	56
Giant Invertebrates of the Palaeozoic Era	57
Giant Amphibians of the Palaeozoic Era	60
End of the Palaeozoic Era	60
Age of the Dinosaurs	61

Mesozoic Era Plants	66
Mesozoic Era Pterosaurs	67
Disappearance of the Largest Dinosaurs	68
Death of the Dinosaurs	70
Super-giant Land Mammals	71
Giant Land Mammals and Birds	72
The Emergence of Mankind	74
Man's Effect on Wildlife	77
The Gradual Scale Reduction of Life	78
The Changing Force of Gravity	80
Evolution into Smaller Sizes	81
3 - Drifting Continents	**83**
The Foundation of Geology	87
The Beginning of a New Theory	88
Continental Drift is Rediscovered	98
The Expanding Earth Theory is Rejected	102
4 - The Expanding Earth	**113**
Confirming the Expanding Earth Theory	115
Ancient Surface Gravity	126
Increasing Radius, Mass and Surface Gravity	130
The Earth of 300 Million Years Ago	131
Uniform Temperatures	131
A Model of Uniform Temperatures	133
The Splitting of the Continents	134
5 - Meteorites and Ice Ages	**139**
New Matter	139
Cosmic Particles	141
Meteorite Bombardment	142
The Timing of Formation	145
Meteorites, Asteroids and Comets	146
The Heavy Metal Layer	151
The Quantity of Material Involved	151
Discs of Dust	153
Checking the Rate of the Earth's Expansion	155
The Transport of Material into the Core of the Earth	156
Isolated Subduction	158
The Ice Ages	161
Cosmic Winters	164

Small Comets	164
Linking Mass Extinctions to Climate	168
Little Ice Ages	169
A Message for the Future	171
6 - The Solar System	173
The Moon	174
Mars	179
Venus	185
The Giant Planets	186
From Brown Dwarfs to Stars	186
Distant Galaxies	187
Milky Way Galaxy	189
The Solar System	191
7 - Ancient Earth	193
Atmosphere	194
Temperature	195
Life	195
Developing Atmosphere	198
First Cellular Life	199
Glaciers and the Snowball Earth	200
First Seas	202
Multi-cellular Life	203
First Life on Land	204
Final Thoughts	205
References	207
Index	213

Figures

Fig. 1.1 Skeleton of Diplodocus Compared to an Elephant   30

Fig. 1.2 Deinonychus as a Bird-like Dinosaur   32

Fig. 1.3 Tanystropheus   33

Fig. 1.4 Scale Effects on Boxes   36

Fig. 1.5 Scale Effects on Bridges   37

Fig. 1.6 Optimum Form of Bridges   38

Fig. 1.7 A Small and Large Animal in the Same Gravity   39

Fig. 1.8 Scale Effects on Bones   40

Fig. 1.9 Elephant and Ant Optimum Form   41

Fig. 1.10 Galileo's Scale Effect   43

Fig. 1.11 Silk Cotton Tree   47

Fig. 1.12 Comparing Animals in a Different Gravity   51

Fig. 2.1 'Living Fossil' Coelacanth   54

Fig. 2.2 Ancient Giant Horsetails   57

Fig. 2.3 Giant Dragonfly   58

Fig. 2.4 Giant Millipede   59

Fig. 2.5 Amphibians to Dinosaurs 60

Fig. 2.6 Berlin Brachiosaurus 62

Fig. 2.7 Brachiosaurus and Other Larger Leg Bones 63

Fig. 2.8 Diplodocus's Neck Ligament 64

Fig. 2.9 Stegosaurus 66

Fig. 2.10 Pterosaur 67

Fig. 2.11 Modern Rhinoceros and Triceratops 68

Fig. 2.12 Duck-billed Dinosaurs 69

Fig. 2.13 Uintatherium 71

Fig. 2.14 Giant Land Birds 72

Fig. 2.15 Giant Irish Elk 73

Fig. 2.16 Giant Ground Sloth 73

Fig. 2.17 Giant Armadillo 74

Fig. 2.18 Mammoth 75

Fig. 2.19 Sabre-tooth Cat 76

Fig. 2.20 Largest Life Over Geological Time 79

Fig. 2.21 Relative Scale Reduction of Life 80

Fig. 2.22 Gravity Increase over Geological Time 81

Fig. 3.1 Snider-Pellegrini and Baker Reconstructions 89

Fig. 3.2 Continental Blocks 'Float' on the Denser Mantle 91

Fig. 3.3 Wegener Reconstructions 93

Fig. 3.4 Hilgenberg Expanding Earth Reconstructions 95

Fig. 3.5 Laurasia and Gondwana Reconstructions   96

Fig. 3.6 Convection Cell   103

Fig. 3.7 Intermittent Subduction Zones   104

Fig. 3.8 Gaps on an ancient Constant Diameter Earth   106

Fig. 4.1 Extension and Compression of the Earth's Crust   116

Fig. 4.2 The Author's Expanding Earth Reconstructions   118

Fig. 4.3 Earth's Changing Radius from Geological Data   126

Fig. 4.4 Gravity Variation of Known Celestial Bodies   128

Fig. 4.5 Earth's Changing Gravity From Geological Data   129

Fig. 4.6 Changing Climatic Belts   133

Fig. 5.1 Earth Formation Theories   145

Fig. 5.2 Eros Asteroid   147

Fig. 5.3 Comet Impact   148

Fig. 5.4 Impact Crater   149

Fig. 5.5 Horsehead Nebula   152

Fig. 5.6 Sombrero Hat Galaxy   153

Fig. 5.7 Cosmic Seasons   154

Fig. 5.8 Stalactites and Stalagmites   157

Fig. 5.9 Mass Transport into Earth   159

Fig. 5.10 Spiral Galaxy M51   162

Fig. 5.11 Frost Fairs   170

Fig. 6.1 Comparison of Planetary Expansion   174

Fig. 6.2 Valley of the Alps   175

Fig. 6.3 Ariadaeus Rill   176

Fig. 6.4 Moon Footprint   177

Fig. 6.5 Tycho Crater   178

Fig. 6.6 Valles Marineris on Mars   180

Fig. 6.7 Arandas Crater   181

Fig. 6.8 Martian Surface   183

Fig. 6.9 Map of Venus   185

Fig. 7.1 Earth's Developing Atmosphere   195

Fig. 7.2 Stromatolite   197

Fig. 7.3 Gas Relative Weights   199

Fig. 7.4 Division of Cellular Life   200

Introduction

The dinosaurs have been a source of wonder and fascination since they were first discovered. A large part of this fascination is their gigantic size, since they were the largest land animals ever to live with most average-size dinosaurs dwarfing the largest land animals of today. Although the gigantic size of dinosaurs is obvious, the reason has remained a mystery for over a century. Why were the dinosaurs so huge? What was so different about the world in those ancient times?

In October 1987, while on a lazy beach holiday in Portugal with my wife and son, I pondered on this question of the dinosaurs' gigantic size compared with present-day life. As a design engineer, I was particularly interested in calculations which showed that the bones of the larger dinosaurs were too weak to support their own body weight. Here was the essential paradox of the dinosaurs' large size. Their bones should buckle and crack. Yet the fossil bones in museums around the world showed that these giants had thrived in their own world of hundreds of millions of years ago. How can both of these statements be true? How is it possible for the dinosaurs to dwarf the life of today?

There is one simple, yet astonishing, answer. Today's life has evolved to live in our present gravity. In a reduced gravity dinosaurs would weigh less - so bones, ligaments and muscles could be weaker. Blood pressure would also be less. Effectively, the scale of life is controlled by gravity, so a weaker gravity would allow *all* life to become larger. Dinosaurs could become huge if the ancient Earth's surface gravity was weaker than the present gravity.

The explanation is beautifully simple in its clarification of the dinosaurs' gigantic size. Using the idea of an ancient Reduced Gravity Earth allows a fascinating new world in which the

Introduction

animals of the past grew to gigantic proportions in a reduced gravity.

Over millions of years of geological time since the dinosaurs, the relative scale of life should reduce as gravity increased to the present-day value. There can be no doubt that this scale reduction of life has taken place. After the dinosaurs became extinct, a range of super-giant mammals reached the size of the smaller dinosaurs. Millions of years after them came giant versions of the animals of today. These died out within the last few million years to leave their smaller present-day cousins.

One possible reason for a Reduced Gravity Earth was that the ancient Earth was smaller with a reduced gravity to match. A few simple calculations on the Portugal beach in 1987 quickly revealed that the ancient Earth would need to be about half its present diameter to account for the scale of life during the dinosaurs' time. But just like most people I had taken it for granted from an early age that the dinosaurs' Earth was the same diameter as today's Earth. It seemed impossible that the ancient Earth had such a small diameter.

All these doubts were rapidly overturned when I returned home and researched the local library (there was no Internet in those days). It was soon obvious that several geologists had already proposed such a smaller diameter ancient Earth based on geological evidence that the Earth had expanded to its present size - an Expanding Earth.

Could it be true? Could it be so simple?

During the following years I contacted various geologists to help promote this concept. The British geologist Hugh G. Owen, who at that time was working at the London Natural History Museum, suggested that an easier method of promoting the significance of this concept more widely was to write a book. After his prompting the first edition of *Dinosaurs and the Expanding Earth* was finally published in 1994.[1]

The book was an initial attempt to widely publicise the evidence indicating an ancient Reduced Gravity Earth. In that first edition I asked readers to contact me with any evidence that would throw light on the thesis of dinosaurs living in a reduced gravity. The invitation generated hundreds of responses from people all around the world and there have now been many interesting observations.

The Reduced Gravity Earth theory created a lot of controversy that is generally polarised into two widely different viewpoints.

---
[1] *Hurrell, 1994.*

Dinosaurs and the Expanding Earth

Some people were (and still are) adamant that gravity was not less on the ancient Earth. A few were so convinced they were correct they seemed annoyed with me for even daring to suggest the possibility. But our scientific understanding of the world is still evolving, so we should be wary of insisting we are 100% right about everything we think we know. I take the more open-minded view that some of what we think we know could be wrong. This view of science is supported by history - look back 100 years and some of what was being taught as scientific fact was different to today. Look back another 100 years and the scientific truth of the day was different again. There is still room for radical new ideas.

Fortunately there were also other more open-minded people who were prepared to consider the possibility of a Reduced Gravity Earth. After reading a review of *Dinosaurs and the Expanding Earth* in the science magazine *Geology Today*, the Australian geologist John K. Davidson read my book in 1996 and was so impressed with the implications of the concept that he passed his own copy onto his former geology professor and obtained another copy of the book for himself.[1,2]

Davidson's former professor was one of the leading advocates of the Expanding Earth theory - Sam Warren Carey, a Tasmanian Professor of Geology who championed the Expanding Earth theory for decades. He had long since retired but was still active in promoting the Expanding Earth theory to explain various geological observations that had troubled him for some time. A few years later in 2000 the second edition of Carey's last book, *Earth Universe Cosmos*, now included a new section on dinosaurs in reduced gravity and concluded:

> Mesozoic dinosaurs could not have existed with present surface gravity, nor would have bat-like pterosaurs with 12 metre wing spans. Engineers (Hurrell, 1994) have shown that dinosaurs' bones could not have borne their weight ...
>
> The size of dinosaurs peaked in the Jurassic with *Diplodocus*, *Brontosaurus*, and flying reptiles like *Quetzalcoatlus*. By the mid-Cretaceous *Triceratops* and *Tyrannosaurus rex* were much smaller, although still huge. Oligocene animals were much smaller although very much larger than their modern relatives. Birds

---

[1] *John K. Davidson personal correspondence, 1996 & 2001.*

[2] *Geology Today, July-August,1996/159.*

Introduction

became lighter from the heavy-boned *Archaeopteryx* and the bird-like *Iguanodon* to much lighter modern birds.[1]

Carey was obviously publicly supporting the concept of a Reduced Gravity Earth and clearly stated that gravity must have been less on the ancient Earth to enable dinosaurs to reach such immense sizes.

Carey was one of the founding fathers of the Expanding Earth theory based on geological evidence and had already contemplated one key unknown about the concept - was the Earth expanding under conditions of constant or increasing mass?

The evidence for a Reduced Gravity Earth theory clearly provided additional unrelated support to the geological evidence for an Increasing Mass Expanding Earth. Both theories reinforce each other and Carey also picked up on this key point:

> Reduced Earth radius with constant Earth mass implies higher surface gravity, but much reduced surface gravity is essential for dinosaurs to have existed. The mass of the Earth must have been less.[1]

As well as presenting Professor Carey with a copy of my book Davidson also had an astonishing piece of fresh evidence for a Reduced Gravity Earth.[2] He directed me to a scientific paper, published by two American professors in the Geological Society of America magazine, *Geology*, that presented remarkable direct evidence indicating that the Earth's gravity had changed since the dinosaurs' time.[3]

Professors C. John Mann and Sherman P. Kanagy had examined the angle of repose in sandstone blocks, which were once ancient sand dunes, and then compared this angle of repose with the present-day value to calculate ancient gravity.

'Angle of repose' is the angle between the horizontal plane and the surface of dry sand forming a hill. As commonly seen when sand dunes form, at steep angles the sand readily slides downhill but as the angle is reduced the sand reaches a point where it eventually stops moving. The angle where the sand stops sliding down is the angle of repose.

The key point of the professors' research is that the angle of repose changed the further back in time they looked. Since the

---

[1] *Carey, 2000 (see page 131).*
[2] *John Davidson personal correspondence, 1996.*
[3] *Mann and Kanagy, 1990.*

angle of repose is directly related to gravity, measuring this angle in ancient sand dunes indicated that:

> ... steeper angles may have been recorded in ancient sediment because Earth's acceleration of gravity was less than now.[2]

It seems that they may have found an ancient record of the changing gravity of the Earth. The whole theory is subject to record and calculation so the reduced gravity of the past can be effectively measured from the angle of repose formed by ancient sand dunes.

Professor Mann confirmed he was still investigating the evidence when I wrote to him. Even though he warned me he was not an advocate of the Expanding Earth, we agreed to keep each other apprised of our own work in this area in view of the fact that we were both continuing to pursue answers to our questions about gravity changes on Earth.[1]

Although Mann and Kanagy did not support the Expanding Earth theory, Davidson had explored the concept that this variation in gravity might be caused by some form of physical change in the Earth. He used the ancient sand dune data to calculate the size of the Earth expansion required to account for the increase in gravity in a scientific paper published in the *Frontiers of Fundamental Physics*.[2] Once again the results supported each other. All this evidence for a Reduced Gravity Earth was becoming very interesting!

Davidson wonders in some respects if the problem has wider implications since increasing gravity and present creation of matter is at the root of all science:

> Increasing gravity evidenced by either biomechanics and the maximum angle of repose of sands has to be considered by all three possible Earth radii; contracting, constant and expanding. A contracting radius requires a starting radius, some 250 million years ago, approximately 50% greater than today's if the mass is kept constant. This would result in a residual present Pacific, after absorbing the Atlantic and Indian Oceans, about five times greater than it is. Either a constant radius or expanded earth requires increasing mass and an answer to the fundamental question, where did the

---

[1] *C. John Mann personal correspondence, 1997.*

[2] *Davidson, 1994.*

Introduction

> mass come from? (Your book seems) to make the increasing mass problem principally one for the Earth Expanders alone, but the constant radius Plate Tectonicists have a lesser, but identical problem. This is the fundamental challenge to be addressed by all of us, especially the particle physicists.[1]

Another correspondent was the Australian geologist James Maxlow who wrote in 1998 that:

> Your dinosaur information cleared up a very contentious issue for me regarding whether the Earth is expanding under conditions of constant mass, or increasing mass. ... I checked my mathematical modeling today and discovered that under conditions of mass increase the surface gravity ... was ... precisely what you are suggesting. Prior to this I had been erring towards a constant mass scenario because of the, what I thought was an, unacceptable increase in mass for the future.[2]

Maxlow later published a book in 2005 about the Expanding Earth, *Terra Non Firma Earth*, which included a short section on dinosaurs and other prehistoric life and stated in part:

> It can be seen that on an Expanding Earth, surface gravity during the Precambrian Eras would be about one third of the present value and about one half of the present value during the Mesozoic Era. The Mesozoic Era of course was the Era of dinosaurs, those very large, very long bodied creatures who could very well have benefited from a much lower surface gravity.
>
> ... for an Earth undergoing expansion as a result of an increase in mass over time, the surface gravity during the Triassic Period would have been approximately 50 percent of the present value. This then increased to approximately 75 percent of the present value during the Late Cretaceous Period.
>
> Considering the large size and length of many of the dinosaur species, this much reduced surface gravity would have benefited their existence and mobility immensely. The progressive increase in surface gravity over time may then offer an additional explanation for

---

[1] *John Davidson personal correspondence, 2010.*
[2] *James Maxlow personal correspondence 1998.*

the relatively rapid turnover of dinosaur species throughout their long history.[1]

Some people wrote that they had similar ideas about dinosaurs living in reduced gravity and a few of these have published their own thoughts on the subject in various forms.

Ramin Amir Mardfar is an Iranian who is interested in the evolution of animals and the effect of the Earth's gravity on it. In the year 2000 he published his book, *The relationship between Earth gravity and Evolution,* which comprised 13 of his scientific papers previously published in *Etelaat-e-Elmi Magazine.*[2] This book was written in Farsi, the official language of Iran, but fortunately for those of us who don't speak this language he placed an English translation on his web site.

Mardfar also discussed scale effects relative to an animal's blood circulation system. Fish have the simplest blood circulation with two-chamber hearts. Reptiles have three-chamber hearts. Mammals and birds have four-chamber hearts. These improvements in blood circulation can all be related to the higher blood pressure required to live at a larger scale.

William Carnell Erickson has been interested in reduced gravity for many years and published an article in 2001, *On the Origin of Dinosaurs and Mammals.* Erickson argued:

> ... that natural selection in reduced gravity will favor bone thinning, a relative decrease in skeletal mass, and an increase in the uppermost limit to body size. These predictions are borne out in the fossil record: the Late Triassic witnessed the proliferation of gracile, long-limbed and lightly-constructed diapsid reptiles (thecodonts and dinosaurs) at the expense of the synapsid (mammal-like) reptiles, animals that were much more compact, cumbersome and massively-constructed. Giant dinosaurs, such as *Melanosaurus*, were already present in the Late Triassic, followed soon thereafter by the largest of all land-living animals, the sauropods.[3]

Lance Endersbee, who was Emeritus Professor of Civil Engineering at Monash University, had also followed the debate as part of his interest in geology and commented in 2000 about

---

[1] *Maxlow, 2005.*
[2] *Mardfar, 2000.*
[3] *Erickson, 2001.*

Introduction

the Reduced Gravity Earth theory in an article in the science journal, *ATMS Focus:*

> ... the force of gravity may have been less in earlier times. Stephen Hurrell, a design engineer, worked out that the bones of the larger dinosaurs were too weak to support their own body weight. He concluded that they could only have evolved at a time when the force of gravity at the Earth's surface was much less than at present.[1]

The Italian geoscientist Giancarlo Scalera produced a compilation of some of the main arguments for a Reduced Gravity Earth in the 2003 book *Why Expanding Earth?,* which was dedicated to Ott Christoph Hilgenberg, one early pioneer into the concept of the Expanding Earth.[2] Scalera's review noted:

> The possibility that paleogravity could be lesser and lesser going back through geological time has been defended on different bases. The first topic says that the Dinosaur sizes and their bone architecture (especially the giant biped Dynos) are not suitable to walk and still more difficult, to run ...
>
> From the vegetal realm it is possible to note a progressive decrease of the dimensions of the trees, whose maximum height depends on the possibility to transfer water and sap to higher leaves, and then directly from gravity. ...plants, from angiosperm to gymnosperm and so on have decreased in size on the average through Mesozoic and Cenozoic, and this can be undoubtedly interpreted as evidence of lesser Mesozoic gravity.[3]

A more comprehensive scientific examination of the implications behind any possible gravity variation is contained in Scalera's paper *Gravity and Expanding Earth.*[4] He has also reviewed the first two editions of *Dinosaurs and the Expanding Earth* and concluded in part:

> Have you seen the huge dinosaur's footprint in Colorado? Or in Australia? Have you looked critically at a dinosaur skeleton in a Paleontology Museum ... ?

---

[1] *Endersbee, 2000*

[2] *Scalera and Jacob, 2003*

[3] *Scalera, 2003.*

[4] *Scalera, 2002 & 2004.*

> Stephen Hurrell's book ... offers a discussion of this problem, proposing an increasing gravity throughout geologic time. ...[1]

The Internet has spawned numerous web sites which promote the Reduced Gravity Earth theory and various people are also arguing the case for a Reduced Gravity Earth in Internet discussion groups like *Unexplained-Mysteries.com* and *The Paleogravity Chat Room*.

One of the most noticeable people using the Internet to argue the case for a Reduced Gravity Earth due to an Increasing Mass Expanding Earth is Neal Adams who has also been interviewed on American Radio explaining why he believes dinosaurs lived in a reduced gravity. As he explained to Art Bell on the American Coast to Coast radio program in 2005, and a number of other radio shows:

> There is a debate on the Discovery Channel between two very prominent paleontologists who talk about a *Tyrannosaurus rex* and they say it was the size of, or larger than, an elephant. One of them says it was a scavenger because it could not run faster than 10 miles per hour. The other scientist - a very profound scientist - says no, that's not true. It was a predator: it chased down and killed its prey. The other scientist says it could not chase down and kill its prey because if it took a right turn going 50 miles per hour its head would snap off. Well, you know, it's true. Its head would be the weight of a motor cycle and bone really can't carry that kind of weight. ...
>
> This debate can easily be settled by assuming that gravity (was less than today).[2]

The book sales of *Dinosaurs and the Expanding Earth* have not followed the typical pattern of most books where sales generally decrease after the initial surge of interest. Sales of the book have gradually increased over the years as the Reduced Gravity Earth theory began to be more widely known and debated. The first hard back edition sold in small but steady numbers so it needed to be reprinted in 1996 and 2001. An ebook edition was also released in 2003. But as the theory and its implications became more widely known, sales gradually increased year on year and

---
[1] *Scalera, 2006.*
[2] *Neal Adams's interview on American Coast to Coast Radio Program, 2005.*

Introduction

various people began asking me for another paper copy of the book to be reprinted.

Rather than just allowing the book to be reprinted, I have taken the opportunity to make some minor edits and explain various points in greater detail, as suggested by many readers. I hope this has made this third edition an even better and more interesting book and must pass on my sincere thanks to everyone who has suggested improvements.

Acceptance of the possibility of a Reduced Gravity Earth has gained ground slowly but steadily over the years since the first edition of *Dinosaurs and the Expanding Earth* introduced the theory. Some disregard the new concept without much thought but others think deeply about it and see the merits of the idea. These individuals tend to pass the information onto others so it gradually becomes more widely known and debated. Today there are some notable supporters of the Reduced Gravity Earth theory.

This book can only be a brief introduction to the concept. The Reduced Gravity Earth theory, and its implications for the Expanding Earth, will require years of further open-minded research and support to develop and refine the concepts. It needs your support.

1 - The Scale Limits of Life

Intense study of prehistoric animals has demonstrated beyond doubt that the dinosaurs were the largest land creatures which ever lived. Their immense size has fired man's imagination since their first discovery - even the name dinosaur means terrible lizard. This gigantic scale has often been a cause of wonder to many people, from young to old, thereby ensuring that the continuing fascination with dinosaurs has the one main question: *Why* were the dinosaurs so large?

The dinosaurs which have the honour of being the most gigantic four-legged land animals to have ever lived are the sauropods. They had long necks and tails with four elephant-like legs, and they included among the more well-known, *Apatosaurus*, (formerly known as *Brontosaurus*), *Brachiosaurus* and *Diplodocus*, as shown in Figure 1.1. Some achieved incredible sizes. Many measured from 20-25 metres (65-82 ft) long and weighed over 50 tonnes. This is several times the weight of today's largest land animal, the African bull elephant, which weighs 7-8 tonnes on average.

In 1912 the largest complete skeleton of any land animal was discovered in East Africa and shipped to Berlin Natural History Museum for display. The mounted reconstruction of the Berlin *Brachiosaurus brancai*[1] is 22.5 metres (74 ft) long, stands 12 metres (40 ft) high, and it has been estimated to have a mass as great as 12 bull elephants. The shape of its deep, narrow chest and its legs suggest an animal which lived on land as a gigantic giraffe-like animal, browsing the tops of trees.

The immense size of the dinosaurs has given problems from their earliest discovery. When a bipedal dinosaur, *Iguanodon*, was first described its size was so massive that it was

---

[1] *It has recently been suggested that Brachiosaurus brancai should be reclassified as a separate genus Giraffatitan brancai but the original Brachiosaurus brancai classification is used in this book.*

Chapter 1 - The Size Limits of Life

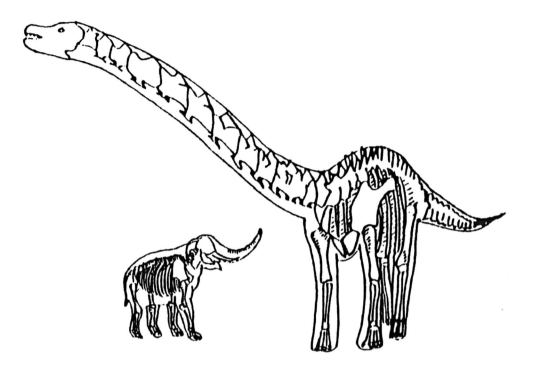

Fig. 1.1 Skeleton of Diplodocus Compared to an Elephant

The skeleton and outline of one of the sauropods, Diplodocus, and the Indian Elephant, both drawn to the same scale, show the large scale of the dinosaurs. Diplodocus stood 3.8 metres (12.5 ft) tall at the hips. The Indian Elephant is 3.1 metres (10.2 ft) head height and the African Elephant is 3.3 metres (10.8 ft) head height.

reconstructed in London's Crystal Palace Park, in 1853, as a stout-legged, four-footed dinosaur. This was hardly surprising as *Iguanodon*'s mass was greater than today's elephant and it was over 10 metres (33 ft) long. After more than a century of further study, modern reconstruction has confidently shown that *Iguanodon* was only a comparatively small dinosaur that walked on two legs with a remarkably slender form that looked more like a giant flightless bird.

As the gigantic scale obtained by dinosaurs is so huge it is difficult to believe that any present-day animal could achieve the same scale. Calculations show that the leg bones of the larger dinosaurs would tend to break under their own weight. This leads to the conclusion that they could not exist in today's world - and yet they certainly did exist in the world of 150 million years ago. Because of this, the dinosaurs' gigantic proportions must be recognised as one of the most intriguing puzzles of our time.

The Paradox of Scale and Lifestyle

The giants of the past have presented palaeontologists with a paradox that could make your head spin. One side of the argument shows that the dinosaurs were too large to move fast without injuring themselves, but on the other hand detailed reconstruction gives the impression that they were agile, active creatures like the smaller animals of today. So which is true? It's the sort of problem which seems impossible to solve. This paradox has resulted in two completely different views of dinosaurs. The dinosaurs were agile, quick and warm-blooded - or the dinosaurs were slow, clumsy and cold-blooded.

These two radically different views have alternated with each other as the established view. The first dinosaurs of the early 1800s were reconstructed as stout four-legged animals, until further studies of the arrangement of their legs and muscles towards the turn of the century showed that these animals must have been agile and fast. In Britain Richard Owen and Thomas Huxley described how various dinosaur characteristics were bird-like. Over in America Edward Cope and Charles Marsh made similar connections between dinosaurs and birds. The consensus among the best palaeontologists of the time was that the dinosaurs were the direct ancestors of the birds of today - and since the birds of today were agile the dinosaurs must have also been relatively agile.

This view was generally accepted until the First World War, but between the two world wars the dinosaurs again became slow and lumbering as the arguments for their relationship to birds were forgotten.

During the 1970s, the connection between birds and dinosaurs began to be re-established. In America, John Ostrom, a Professor of Palaeontology at Yale University in the USA, spent two years analysing the meat-eating dinosaur *Deinonychus*. His bio-mechanical analysis showed that it must have had high levels of manoeuvrability and stamina so it was very bird-like, as shown in Figure 1.2. Later, he took the connection between

Chapter 1 - The Size Limits of Life

Fig. 1.2 Deinonychus as a Bird-like Dinosaur
Analysis of the meat-eating dinosaur, Deinonychus, indicated that it was a highly active animal which was very bird-like.

dinosaurs and birds further when his earlier studies allowed him to see the relationship between the fossils of the dinosaur *Deinonychus* and the oldest known bird, *Archaeopteryx*. Nearly every detail of the finger, shoulder, hip, thigh and ankle of the two animals was identical. The connection between dinosaurs and birds was again firmly established.

With birds as the direct descendants of dinosaurs it becomes impossible to see them as slow lumbering creatures. The metabolic rate of birds is higher than mammals so the dinosaurs must be reconstructed as fast and mobile animals capable of dominating the land despite their large scale.

Perhaps the best description of what these dinosaurs might really have been like comes from one of the world's leading palaeontologists, Robert T. Bakker. In his book, *The Dinosaur Heresies,* he imagined dinosaurs living as agile hot-blooded animals.[1] He describes sauropods capable of stretching their long necks to reach the tops of conifers while standing on their hind legs. They might even have reared up on their hind legs to defend themselves against the carnivorous dinosaurs of their time. When walking or running, their long tails were held erect to counterbalance the weight of their equally long necks.

Among the hunters of the dinosaurs, he believes that *Tyrannosaurus rex*, perhaps the most famous of all dinosaurs, was a fast and agile hunter. The legs were built for speed with massive muscles capable of propelling *Tyrannosaurus* forward at great speed. The lung and heart cavities were equally large to enable him to pump the blood and oxygen required by his massive leg muscles.

---

[1] *Bakker, 1986.*

Long Necks

Although the dinosaurs are now again being reconstructed as dynamic creatures it still leaves the paradox of their exceptionally large scale. Often the one animal can be reconstructed using these two differing views to obtain conflicting descriptions of the same animal.

A good example comes from the extensive studies of the fossil animal *Tanystropheus* found in the middle Triassic sediments of Monte San Giorgio in Switzerland. These studies have resulted in two palaeontologists having completely opposite views of how these animals lived.[1] The sediments now form part of a mountain in the Alps, but 230 million years ago they were being deposited in a small inland sea. *Tanystropheus* lived on the fish and cephalopods in this sea and used its webbed hind feet to swim in the water. But it must have also been able to walk on land to lay its eggs out of water.

The biggest problem with reconstructing how *Tanystropheus* lived is its long neck. The long neck could have been flexible and used for catching prey as the animal floated on the water or lay on the shore. When walking on land the neck could have been held rather like a modern swan's, in an 'S' curve, as illustrated in Figure 1.3, but the neck muscles seem to be too weak for such a lifestyle. The shallow vertebrae provided little area for

[1] *Wild, 1980, 1987 & Tschanz, 1985.*

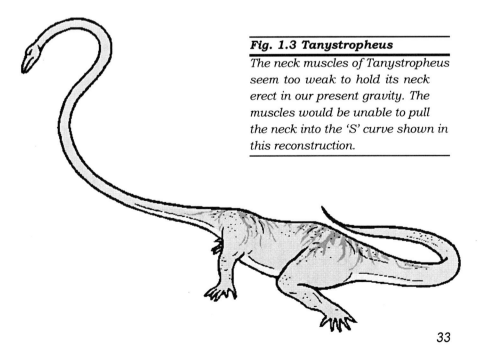

Fig. 1.3 Tanystropheus

The neck muscles of Tanystropheus seem too weak to hold its neck erect in our present gravity. The muscles would be unable to pull the neck into the 'S' curve shown in this reconstruction.

Chapter 1 - The Size Limits of Life

attachment for the neck muscles so these seem to be too weak to pull the neck into an 'S' curve against gravity.

If the neck muscles were too weak to hold its neck erect then the only other alternative is that *Tanystropheus* rested its neck on the ground when it came onto land. But if this is the case then the long neck would be more of a hindrance than a help. Can we really imagine an animal which could not raise its head off the ground? The problem all revolves around the fact that this animal seems too large and heavy for the lifestyle it must have followed.

Previous Attempts to Solve the Mystery

There have been many attempts to explain why prehistoric life forms, and in particular the dinosaurs, have been so gigantic. The results from the fossil records have given such a diverse, and often confusing, set of facts that no one theory has yet been accepted as true. Almost every book written about the dinosaurs explains their huge scale from the writer's personal view and each newly discovered fact has apparently generated a new theory.

Some of these theories have considered only one aspect of the question. For example, a view that was popular until the 1960s was that the sauropods had supported their massive bulk in water. With most of their bodies immersed in water they only raised their long necks and heads out of the water to feed. This view caused much discussion. On one side of the argument was the fact that these animals appeared to be too huge to support themselves without the aid of water. On the other side of the argument, it was pointed out that if these animals had immersed themselves in water, the weight of water would have crushed their lungs, preventing them from breathing.

Footprints which helped to change this view were discovered in the Palurg River, Texas, along with several others in the 1970s. Embedded within the fossilised mud of these footprints are water creatures which only live in shallow water. The footprints prove that these huge dinosaurs, with feet nearly a metre (3 ft) across, walked in this shallow water and could just as easily have walked on land.

It is this type of theory which can never provide a complete answer to the general gigantic scale obtained by prehistoric life. Even if this one group of dinosaurs had been able to grow huge by living in water, it does not explain why so many of the other forms of life were also gigantic. The dinosaurs have become famous for their great scale, but many different forms of life achieved their most gigantic sizes during the Mesozoic Era. The fossil record also shows giant plants, insects and sea creatures. All forms of life seemed able to grow 2-3 times the scale and 4-9 times the mass of the equivalent life of today.

No satisfactory explanation has yet been given for this ability of life to achieve these gigantic sizes, even though this fact has become increasingly obvious over the last century. Only a world-wide effect provides a complete answer and this has lead to ideas linking the giant scale of life with a global mass extinction which wipes out large proportions of life, especially the large life, at regular intervals. A favourite theory is a major catastrophe which killed off all the large animals but left the smaller ones to survive. These catastrophes include such things as a sudden reduction, or rise, in the Earth's temperature, gamma rays from an exploding star in outer space, or giant meteors hitting the Earth. But these catastrophe theories have difficulty explaining how large animals were affected and the small animals were not.

Any believable theory which seeks to explain the scale reduction of life must describe some world-wide force that has changed over hundreds of millions of years and affected all life. Some previously described theories seem to come closer to answering this question than the major catastrophe theories. One theory considered that warmer climates allowed life to become larger in prehistoric times, and a gradual cooling of the climate over millions of years reduced the scale of life. Yet existing life in cold regions is large compared to tropical forms, taking for example polar bears, walruses and giant redwoods. Also, today's giants of the warm regions are no comparison in scale with the giants of the past.

Another theory is that competition from small animals killed off the large ones. This doesn't agree with the known record of life. The fossil record shows that once the large animals died out the remaining smaller animals slowly took over their position as the giants of the time, although they never match the scale of their ancestors.

Other theories can be compared with the evidence in similar ways but they all leave us with one main unanswered question - why did the small animals that survived the mass extinction never become as large as their predecessors? Even if we accept without doubt that some form of major catastrophe kills all large

Chapter 1 - The Size Limits of Life

life forms every 30 million years or so, it does not explain why these small animals never became huge. Life fills every possible size in the struggle to survive - giants would exist in our world if they could. Any theory accounting for the scale reduction of life must describe why this same scale reduction has occurred in all land-based life whenever global changes have taken place.

The Size Limits on Today's Life

The reason all life has reduced in relative scale over a period of hundreds of millions of years has been elusive, the reason seeming so remote in time that it can never be found. Yet a vital clue lies all around us - virtually staring us in the face. It lies in all the animals of today. It starts by asking: Why don't present-day animals grow to the immense proportions of the dinosaurs?

Since we are seeking to explain why the dinosaurs grew to such gigantic sizes we first need an understanding of the factors which govern the maximum scale of today's life. The problem of the giants of the past can be simplified by asking related questions about the scale of modern life: Why is any living organism limited in its size? Why don't elephants grow four times as large as they do? Why don't ants grow to the size of elephants? To explain these we need some factor which can be proven to limit the size of life.

Close study of the life of today should give one clear reason why it cannot achieve the scale of its ancestors. Once we have understood the size limits on today's life, it should be possible to understand why prehistoric life could become so large.

The Effects of Scale on all Structures

Certain fundamental relationships govern the scale of any structure. These mathematical relationships impose an almost

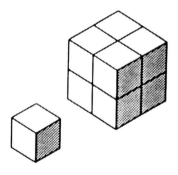

Fig. 1.4 Scale Effects on Boxes

Is the large box 2, 4 or 8 times bigger than the small box? Or all three? Although it seems a trivial question, understanding the answer explains the mystery of scale.

Dinosaurs and the Expanding Earth

unseen order on our universe by acting on all structures to control the scale of both man's and nature's creations. Once we understand these scale effects, as any good engineer or architect must, we can begin to understand the scale effects on life.

Consider the two boxes shown in Figure 1.4. A seemingly simple question is how much bigger is the larger box than the smaller one, but consider the question in more detail. If the boxes are measured by the length of their sides, their linear dimensions, the larger box is twice the scale of the smaller box. If, however, the boxes are measured by the area of their sides, then the larger box is four times the scale of the smaller box. If the boxes' volume is used as the measurement, then the larger box is eight times the scale of the smaller one.

To put it more simply, as any object increases in size its volume increases quicker than its area, and its area increases quicker than its length. It is this mathematical relationship between length, area and volume which limits the maximum scale of any structure. With an understanding of these scale effects, we can clearly understand the relationship between the form and scale of life.

The maximum size of mankind's bridges, skyscrapers and aeroplanes, as well as the forms of life on the Earth, are strictly limited by changing forces due to scale. With the various structures that people build, the designer cannot simply copy an existing design of a different scale. As the scale of a design is increased its form must also change. At larger sizes the effects of weight become more pronounced, and a stronger form of design or material must be used. The largest of these structures have reached the upper limit of their scale. Hence, the development of larger aeroplanes, longer bridges, or taller skyscrapers must await the development of stronger materials.

As one example of the effects of scale on mankind's creations, consider a bridge which uses an arch for support. The basic design of an arch has been used for thousands of years in

Fig. 1.5 Scale Effects on Bridges

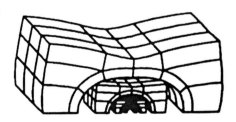

All these bridges are identical in form and are constructed from the same material, but are different sizes. Note how the largest bridge is deformed by its own weight. This is the effect of scale on bridges of different sizes.

Chapter 1 - The Size Limits of Life

Fig. 1.6 Optimum Form of Bridges
Different materials and designs are used for different sizes of structures to allow for the effects of scale. The need to use more complex designs with increasing size changes the 'optimum form' of the bridge. The top drawing opposite shows three bridges drawn to scale. The largest bridge is a complex lattice steel structure. The middle drawing shows an enlarged detail of the two smaller bridges. At this scale a steel bridge design can be simple enough to only require large holes to reduce the bridge weight. The bottom drawing shows the smallest bridge detail. At this scale the bridge can be a simple design of brick with no weight reduction elements. All these design variations are caused by the effect of scale.

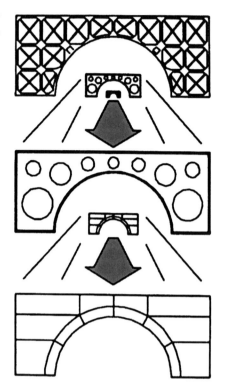

various constructions of bridges. The Romans built many of these, some of which are still standing. Many of the first designs were built in stone, but they were limited in size since over a certain size the arch began to collapse as shown in Figure 1.5. This effectively imposes an upper limit on the size of stone bridges. The only method of building a larger bridge was to use stronger material.

The first iron bridge was built in 1779 at Stonebridge Gorge in Shropshire, England. The use of this new stronger material was an important step towards longer, bigger bridges. It led the way for all the bridges which have been built since then. As even stronger materials were developed the bridges could use longer spans. All these bridges had the same limit on size as the ancient stone bridges - build them too large and they will collapse and this results in different designs and materials for larger scale as shown in Figure 1.6. This idea of a size limit applies to all structures, not just bridges. It also applies to life.

Dinosaurs and the Expanding Earth

Life's Changing Form due to Scale Effects

Let us assume, for the sake of argument, that we have a standard four-legged animal as shown in Figure 1.7. This animal comes in two sizes - small and large - with the weight of each animal supported on its four legs. The linear dimensions (its height, width and breadth) of the large-scale animal are twice those of the small animal. By calculation, the volume, and hence the weight, of the large animal is 8 times the small animal's weight. But the legs of the large animal are only 4 times the area, and hence the strength, of the small animal. The large animal's leg stress is twice the small animal's leg stress because

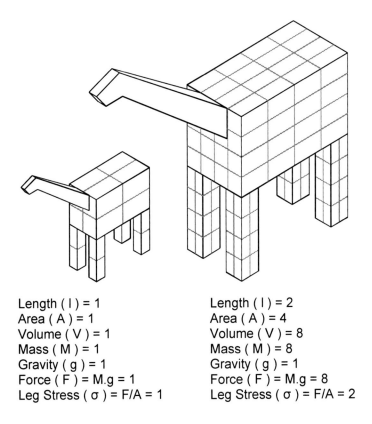

Length (l) = 1　　　　　　　Length (l) = 2
Area (A) = 1　　　　　　　　Area (A) = 4
Volume (V) = 1　　　　　　　Volume (V) = 8
Mass (M) = 1　　　　　　　　Mass (M) = 8
Gravity (g) = 1　　　　　　　Gravity (g) = 1
Force (F) = M.g = 1　　　　　Force (F) = M.g = 8
Leg Stress (σ) = F/A = 1　　　Leg Stress (σ) = F/A = 2

Fig. 1.7 A Small and Large Animal in the Same Gravity

Consider two animals of exactly the same shape, with one animal twice as large as the other, both in the same gravity. The larger animal would have double the leg stress due to the effect of scale. This scale effect sets a boundary to the upper scale of all life.

39

Chapter 1 - The Size Limits of Life

the weight of the large animal has increased quicker than its strength due to the scale effect.

So in practice, if any large animal were twice the scale of the same form of small animal it would only be half as strong, since the stress in the large animal's legs is twice the stress of the smaller animal. Its strength-to-weight ratio would be smaller. This simple example clearly explains why all animals must be limited in size.

But what happens in real life? To overcome this shortfall in strength, the legs of real large-scale animals generally tend to be proportionally thicker for greater strength. But as the animal's scale is increased its thicker legs reduce its mobility. All animals reach an upper size limit when the advantage of larger scale is overcome by the disadvantage of thicker legs.

To counteract the effect of increasing scale the bones of animals tend to become thicker and therefore stronger with increasing scale. The proportion of the weight of an animal's skeleton compared to its total body weight shows how the whole skeleton becomes stronger and heavier with increasing scale. Bones comprise about 8% of the body weight of the mouse or wren, about 14% of a goose or dog, and about 18% of the body of a man.

The relative thickening of an animal's legs, due to the scale effect, can be seen in nature. Take for comparison the thigh bone of a deer, a rhinoceros and an elephant, as shown in Figure 1.8. As the animals increase in weight the relative increase in the thickness of their legs is greater compared to their body size. The deer has the most slender legs, the rhinoceros relatively thicker, and the elephant has great thick sturdy legs to support its massive bulk.

The reason for the unequal variation between an animal's body weight and strength with scale is similar to the two boxes. As any animal increases in scale its volume, and hence its

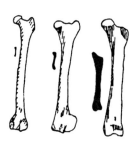

Fig. 1.8 Scale Effects on Bones

In practice the leg bones of larger animals tend to become thicker with increasing size. The silhouettes indicate the true scale of the leg bones and the sketches show the shape of the bones. Note how the larger leg bones become thicker.

Fig. 1.9 Elephant and Ant Optimum Form

The legs of the Elephant and the Ant are the shape they are because of their scale.

weight, increase faster than its strength. This fact imposes an upper limit on body size which applies to all land life. The same basic principles can be applied to land-based animals, plants, and flying animals. As a plant grows in size its weight increases quicker than its strength. As any flying creature grows in size its weight increases quicker than the lifting capacity of its wings. For every form of living creature there is an upper limit to how large it can be.

The effects of scale produce obvious differences between the form of life of different scale. Often, little attention is paid to the scale of animals, but for every scale of animal there is an optimum form, and a change in form is inevitable for a change in scale.

Comparing the body plan of an ant and an elephant, as shown in Figure 1.9, illustrates how the difference in their scale has caused a pronounced difference between the form of the two animals. The ant's legs are thin and stick out sidewards - a very inefficient method of lifting a weight. The elephant's legs are thick and strong, and they are directly under the body - the best position for lifting its heavy weight.

The Optimum Form

The process of evolution forces life to produce the best design since nature does not continue with a failure. Each new adaptation to life will only survive if it can master its environment and over many such adaptations life finds the optimum solution for its own survival. In this way life can

Chapter 1 - The Size Limits of Life

continually adapt and develop optimum solutions to the problem of surviving.

Some of nature's optimum solutions to problems of structural design are so good they have been used by humanity. The architect of the Crystal Palace, Joseph Paxton, based his ideas for its construction on a plant known as the royal water-lily, *Victoria amazonica*. It has large round leaves up to 2 metres (6 ft) in diameter, which achieve great stability from a series of complicated stiffening ribs on the underside of the plant. Each stiffening rib radiates from the centre of the plant becoming thinner as the stresses become lower. All these main ribs are cross-braced by smaller ribs, and these smaller ribs are in turn braced by yet smaller ribs. The whole plant shows the ultimate use of material strength to provide a strong structure with minimum weight.

The design of the Crystal Palace, for the 1851 World Exhibition in London, used the design principles shown in the royal water-lily for a building of gigantic dimensions. It contained as few main struts as possible, with each main strut being cross-braced by a series of smaller struts. Over a hundred and fifty years later, the Crystal Palace is seen as a turning point in architectural history. The main supporting structure was so sparingly used that it gave the impression of lightness.

The inventor of reinforced concrete was a gardener. In 1867, Joseph Menier was trying to make plant tubs out of concrete, but after noting that plants reinforce their weak structures with a stronger framework, he began to insert iron rods within the concrete. This new material allowed the modern large concrete structures of our time to become reality, with massive bridges and multi-storey buildings.

Optimum Form for Different Scales

One of the most famous scientists of all time is considered to be the first to highlight the importance of the scale effect on life. This was Galileo Galilei, the famous Italian mathematician, astronomer, physicist and engineer who is so famous that he is still commonly known by only his Christian name, Galileo, instead of his surname.

Galileo was probably the first scientist to point out that larger animals need relatively thicker bones than smaller animals in his 1638 book *Discourses and Mathematical Demonstrations Relating to Two New Sciences*. He noted that the bones of very large animals must be scaled out of proportion in order to support the weight of the animal. This is the scale effect, and it is one of the most important elements of Galileo's book, together with its implications for living organisms. He made the point that there must be a limit to size since the whole animal cannot be skeleton. He illustrated the point with a drawing of the leg bone of the human body where the drawing, as reproduced in Figure 1.10, shows the monstrous enlargement that would be needed if a person grew to a colossal height. The proportions must change with scale.

Galileo noted that:

> ... if one wishes to maintain in a great giant the same proportion of limb as that found in an ordinary man he must ... admit a diminution of strength in comparison with men of medium stature; for if his height be increased inordinately he will fall and be crushed under his own weight.[1]

Other scientists continued this work. Boyle, (of Boyle's Law) studied the pressure changes in fish; Steno discussed the geometry of muscles. In his book, *On the Movement of Animals*, Giovanni Borelli was the first to calculate the forces in muscles from movement and also compared the heart to a piston pumping the blood around the body. For this work he is considered to be the father of modern biomechanics.[2]

These studies have shown that the process of evolution is effectively the same as choosing a good design - the optimum form. It is good design by nature. This optimum form changes depending on the scale of the creature. For every size of animal there is an optimum form. The structure of a one centimetre animal must be very different from a ten centimetre animal, and the structural strength of larger animals must be stronger still, if their bones are not to break under their own increased weight. The whole of nature's creations use different structures

[1] *Galilei, 1638.*
[2] *Borelli, 1680, 1681.*

Fig. 1.10 Galileo's Scale Effect

Galileo illustrated the problems of scaling a man by showing how a man's thigh bone would need to be scaled in a giant.
© *Galileo Galilei 1638*

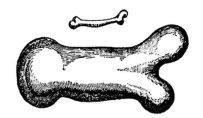

Chapter 1 - The Size Limits of Life

depending on their scale. Small insects need little structural strength but larger mammals require a complex structure of internal bones for strength. The larger animals are more specialised in form because of their size; they can be no other way.

One basic principle of good design is to use the simplest solution to the problem. This is why the insects have come to dominate the region of small size. Insects, because of their very nature, can develop quickly to compete in the struggle for existence, since no resources must be devoted to grow any bones. At this small scale bones would be a positive disadvantage. Hence, no small creatures with bones.

Over a certain scale the optimum form changes and bones are useful. As the scale of animals increases the increasing weight forces the development of bones to withstand the greater stresses of larger size. So two radically different forms of life have evolved to meet two different problems of scale.

These scale effects produce other results. For small animals, the effects of gravity present practically no dangers. Insects like daddy-long legs use fantastic forms of support, while other insects can cling to a ceiling, walk across water by using the effect of surface tension, or float on gentle up-currents of air.

An insect is so small that it can fall without danger. If a mouse falls from a great height, the resistance of the air is sufficiently strong to prevent the mouse reaching any great speed. As it hits the ground it gets a slight shock but can still walk away. With a larger animal, the effects are more severe, with broken bones. Still larger animals are killed by the same fall. These effects come about because as the animal's surface area varies with its own length squared, the same animal's weight varies with its length cubed. Since the surface area controls the animal's wind resistance to falling, the animal's weight increases quicker than its resistance to falling. The larger animals hit the ground faster.

The scale effect also means that small objects will sink slowly in air due to their large surface area-to-weight ratio. This is why dust will stay suspended in air for long periods of time and various plants use the same effect. Plants distribute small grains of pollen over vast areas with a range of up to several hundred kilometres. A grain of birch pollen which was carried up to 2000 metres (6500 ft) by air currents would sink so slowly to the ground that a light breeze would carry it 250 kilometres (155 miles). In good weather grains of pollen have been found at altitudes of 6000 metres (19,000 ft), as high as a jet aircraft.

These grains have such a good surface area-to-weight ratio that they would take over two days to fall to the ground.

Some pollen grains are even more microscopic. The spores of fungus are only 0.005 mm (0.0002 in) in diameter. In free fall from 6000 metres (19,000 ft) they would take about half a month to reach the ground. The pollen of various trees like the horse-chestnut and willow have diameters of up to 0.015 mm (0.0006 in). At the upper limit the pollen grains of the fir tree have a diameter of about 0.15 mm (0.006 in).

At larger scales nature must use new methods to provide a large surface area-to-weight ratio. The dandelion uses a parachute-type structure to transport its seeds on the wind by releasing these light structures on warm air currents. The similarity of the dandelion seeds to a parachute ensures that they cover great distances.

The structure of this parachute seed has been developed in various plants which have found the same solution to the same problem of seed dispersal. There is proof that these plants developed the structures independently of each other because they come from widely different families of life.

The large seed of the maple uses a 'wing' to give a larger surface area-to-weight. As the seed falls from the tree the wing causes the seed to rotate in a spiral path about its own centre of gravity. This allows the seed to fall slowly to the ground and be carried by any currents of air. This seed dispersal method allows the young plants to germinate in new soil away from the shade of their parent plant. Rotating wings have been found to be the optimum form to achieve the slowest rate of descent.

Flight

Exactly the same scale effects occur in flying. The speed needed to keep an aeroplane of a given shape airborne varies with the square root of its length. This is caused by the effect of scale. As any flying object becomes larger its weight increases faster than the lifting capacity of its wings. To increase its lifting capacity it can fly faster, or it can use more efficient wings.

A good example of small scale flight is a paper aeroplane. At small scales the flat wing allows it to fly well, but at the scale of an aeroplane the wing must produce greater lift per surface area by using the aerofoil wing section common to all aeroplanes.

In nature, the smallest of flying insects can grow wings almost as an afterthought. Some insects do not even need wings. One very small lepidopteran moth (*oxyptilus pilosellae*), about 12 mm (0.5 in) long, has wings made up of very fine hairs. Each wing comprises 5 main hairs from which arise about 60 hairs about 0.01 mm (0.0004 in) in diameter and 1 mm (0.04 in) long. At this very small size the aerodynamic efficiency can be very low, and it has been calculated to be only 8% of that required by the dragonfly to become airborne.

Even on the largest of flying insects, the wings are little more than flat sheets. Despite this, insects approach the optimum ideal in flying in that they possess the ability to take off and land vertically - and hover in one place - without losing height.

One of the largest of flying insects is the dragonfly. The dragonfly wing is made up of about 1000 polygons varying in shape and size. The front of the wing has reinforcements running parallel to the leading edge. While the dragonfly is still fast and nimble it has reached the upper limit in size for an insect. The wing is not completely flat, but produces bumps in the surface from one spar to the other. The membranous wings of dragonflies are ultra-light in construction, and yet rigid enough to beat to support flight. This evolved design is so good that the construction and dimensional relationships have been copied by the engineer searching for new methods of construction.

Again because of their small size, small birds are agile with relatively short small wings and quick movements. They can take to the air without any preparation. Because of their similar scale, the stationary hovering flight of the hummingbird and dragonfly are also similar. While the dragonfly has reached the upper limit in scale for insects, the hummingbird has perfected the vertical take-off because it is the smallest of birds. More detailed investigations show that the hummingbird approaches the upper limit of the aerofoil loading of insects.

The weight distribution of birds is critical to their survival, and they must make the optimum use of the weight of their muscles. The smaller birds, such as pigeons, hummingbirds and swallows can take off with ease since their flight muscles can reach up to 40% of their total weight. The relative weight of their leg muscles can be low since their good flying performance allows their legs to be relatively weak. The legs are only suitable for perching, and make up less than 2.5% of the total weight.

Large birds are much less agile with much longer wings in comparison to their bodies. These larger birds need to gain speed along the ground or water before they can fly. Once in the air their movements are much slower than their smaller counterparts. The larger birds such as storks and vultures have relatively weak flight muscles equal to only 12-15% of their total weight. Their legs have very powerful muscles reaching up to as much as 24% of the total weight. These strong legs are necessary for the fast running at take-off. Their scale therefore dictates both their shape and the lifestyle they follow. Smaller birds can hover. A large bird generally uses a rising column of air to soar.

Plants

The optimum form of plants is also affected by their scale. Small weeds rely on the hydrostatic pressure of water within their cells for rigidity. By blowing out their cells with water they form solid structures that can stand erect, even against a strong wind. This use of water to provide rigidity is why a small plant droops as it dries out, but can be revived by placing it in water. The hydrostatic pressure system is adequate to maintain the rigidity of small plants, but as a plant grows larger its weight increases faster than its structural strength.

To cope with the greater stress, larger plants have developed special high strength tissue which we know as wood. The wood fibres of a plant have the sole function of keeping the plant rigid. As the scale increases, plants give way to trees and in large trees we see some of the optimum designs to stop wind blowing the tree over. In the Silk Cotton tree the base of the tree forms buttress roots, as shown in Figure 1.11, to support the canopy of branches and leaves above it. The forces within the trunk

Fig. 1.11 Silk Cotton Tree
The base of the Silk Cotton tree is reinforced with buttress roots to strengthen this large tree. The author is leaning on one of the buttress roots.

Chapter 1 - The Size Limits of Life

increase progressively towards the ground so it is the base of the tree which needs the most reinforcement.

A Solution to the Giants of the Past

By now it should be clear why the scale of prehistoric life is so astonishing - it cannot exist in our world. If someone were to suggest a dragonfly might grow a wingspan of over half a metre (1.6 ft), or that a millipede might grow as long as a cow, or that a horsetail plant, which currently grows to a height of two or three metres (6-10 ft), could grow to 15 metres (49 ft) high, it could easily be shown by calculation that these life forms could never exist at these sizes. The stresses imposed within the living tissues of these large life forms would be too great. Yet although it can be proven that these life forms could not exist in our present world, all these forms have existed at the stated sizes in the distant past. There were giant dragonflies, millipedes and horsetails on the ancient Earth.

The giants of the past therefore demonstrate what appears to be an impossible situation. A perfect paradox. All these giants present the same paradox - they are too large to exist, and yet they did exist on the ancient Earth. Since the giant scale of prehistoric life can be shown to be impossible today surely a responsible world-wide factor can be shown to have changed with time. So what is the answer to the paradox of giant size?

The answer to understanding the puzzle of the gigantic proportions of dinosaurs, and all the other life forms, does not lie in the distant past. The answer lies in every form of life that exists today. By understanding the forces that control the scale of all living things which exist today, we can begin to understand the gigantic proportions of prehistoric life. It is by opening our minds to see the planet's life in a new way, with the mathematical relationship between linear scale and mass, that the reason for the scale of these prehistoric giants becomes clear. The factor which limits various life forms' scale is a combination of scale effects and gravity.

The effect of gravity on the form of today's life has been known for many years. In one well read book written during the First World War *On Growth and Form,* D'Arcy Thompson details the effects of scale on many forms of today's life. He then concludes that:

... the forms as well as the actions of our bodies are entirely conditioned (save for certain exceptions in the case of aquatic animals) by the strength of gravity upon the globe.[1]

He makes no claim to having discovered this fact, but merely describes it as a well-known piece of information (at least among the specialists who make use of it) by recalling the past references of Charles Bell and Crookes to this subject in the 1800s.

Although life has a very narrow scale in which it can exist, it still encompasses the very different conditions in which bacteria, insects and man dominate the environment of their scale. For bacteria the force of gravity does not exist, and its environment consists of the forces of viscosity. For insects the force of gravity can be overcome in the simplest way possible, but the forces of viscosity may trap it within a pool of water. In our world, the force of gravity dominates all the other forces, and our very form is ruled by it.

Having firmly fixed the controlling influence of an animal's body weight on its scale, we can understand that it is the Earth's gravitational field which gives all animals weight and so controls their maximum size. Due to gravity the maximum scale of any structure is limited - this includes all the forms of land-based life. A variation of gravity will change the body weight of all life and will therefore change the maximum scale limit and the optimum form for a certain scale.

Today, there is a whole range of life of similar scale. Among the grazers there are wildebeest, rhinoceros, elephants, deer, and among the carnivores there are lions and tigers. If we now look at the regime of the dinosaurs a totally different scale emerges. There were among the grazers sauropods, with their long necks and tails, and among the carnivores there were bipedal meat-eaters like *Allosaurus*. All these animals were typical of their age and if the whole cross-section of the scale of life at the time of the dinosaurs is compared with today's life then we find that the ancient life was similar in range to today's life, but about 10 times the mass. Where a complete range of plant and meat-eating dinosaurs once roamed, today roam much smaller animals.

This reduction in the scale of life can be explained if we accept that the gravitational field of the ancient Earth, hundreds of

---
[1] *Thompson, 1917.*

Chapter 1 - The Size Limits of Life

millions of years ago, was smaller than the present. The low gravitational field of the distant past allowed the dinosaurs and the other life of that time to reach a much larger scale limit than is now possible. As time passed the Earth's gravity increased and reduced the maximum scale which life could achieve. Today it is the present force of gravity which prevents life growing any larger.

We now have an explanation for the paradox between large scale and lifestyle which has troubled palaeontologists. As I mentioned earlier, their studies had resulted in two different views of how the fossil animal, *Tanystropheus*, must have lived. In a reduced gravity however it could certainly hold its neck upright, even with its typically weak-looking neck muscles. The paradox is resolved with both sides of the argument, those of scale and lifestyle, proven correct.

The Relationship between Scale and Gravity

Having established the factor which limits the scale of life we can begin to ask other questions. How great has this variation of gravity been since the time of the dinosaurs? The first relationship we need to establish, to answer this question, is the relationship between scale and gravity. The simplest way to do this is to examine our standard animal again.

Imagine there are two animals of exactly the same shape, except that the prehistoric one is twice the linear scale of the present-day animal. Under the same gravity, the weight of the prehistoric animal would be 8 times that of the present-day animal while its legs would only be 4 times as strong. The strength-to-weight ratio of the larger animal would be reduced.

This variation can be compensated for by adjusting the strength of gravity as illustrated in Figure 1.12. If the gravitational attraction of the Earth was only one half as strong in the prehistoric time as the present, the prehistoric animal would only be 4 times as heavy and the legs would be 4 times as strong as the present-day animal. Both animals would have the same strength-to-weight ratio because of the difference in gravity.

Both the present-day animal and the prehistoric animal would have reached their optimum scale for their form, but this optimum scale varies according to the gravitational attraction of the Earth. We can now define the mathematical relationship between the optimum scale and gravity:

Length (l) = 1   Length (l) = 2
Area (A) = 1   Area (A) = 4
Volume (V) = 1   Volume (V) = 8
Mass (M) = 1   Mass (M) = 8
Gravity (g) = 1   Gravity (g) = 0.5
Force (F) = M.g = 1   Force (F) = M.g = 4
Leg Stress (σ) = F/A = 1   Leg Stress (σ) = F/A = 1

Fig. 1.12 Comparing Animals in a Different Gravity

Calculating the forces on geometrically similar animals in a different gravity defines the relationship between life's scale and gravity. An animal's leg stress is due to the force of gravity. If gravity is halved then the large animal can double its linear size but its leg stress will still be the same as the small animal's leg stress. This is the controlling influence gravity has on the scale of life - all life can become larger in a reduced gravity in a distinct mathematical relationship.

For a particular form of life the linear scale of land-based life is inversely proportional to the strength of the gravitational field.

This can be represented in a formula as:

$S_r = 1/g_r$

where S_r is the scale of land-based life relative to today's land-based life and g_r is gravity relative to today's gravity.

Chapter 1 - The Size Limits of Life

The effect of gravity on life's scale is a distinct mathematical relationship that affects the basic building blocks of animals - bones, ligaments, muscles and blood pressure. A reduced gravity reduces the force on any animal's bones, ligaments and muscles so they can all be thinner and weaker for a particular scale of life. Blood pressure is also reduced in a weaker gravity since blood pressure is the hydrostatic weight of blood (mass x gravity).

The scale of life is shifted towards a larger scale so dynamically similar animals are larger in a reduced gravity. The most obvious result of this scale shift is gigantic dinosaurs with masses equal to several elephants but the effects are also plain on smaller animals as well. An elephant-sized dinosaur is noticeably more active and dynamic than any elephant because the dinosaur evolved to live in a reduced gravity.

Although this provides an answer to the mystery of the gigantic scale of these ancient life forms, it raises many more questions. What produced this change in gravity? Has this change in gravity been a gradual process operating over millions of years, or has it changed in a series of short jumps, or even one or more large jumps? Is it possible to quantify the scale and rate of change from a study of the life of the past? I will attempt to answer these questions in the next chapters.

2 - The Giants of the Past

We have now considered the relationship between the Earth's gravity and the relative scale of life. The main point which emerged is that a lower gravitational field will allow the scale of all life to become larger, and this illustrates how the immense size of the largest dinosaurs would have been a natural result of a reduced surface gravity on the ancient Earth.

The argument can be extended further. The force of ancient gravity can be estimated from ancient life using a number of different methods including:

- the relative scale of ancient life compared to present-day life,
- ancient life's bone strength,
- muscle strength,
- ligament strength,
- and blood pressure.

One of the easiest methods to estimate the Earth's ancient gravity is to compare the relative scale of ancient life to present-day life since any change in the relative scale of prehistoric life over time quantifies how the Earth's gravity has also changed.

As discussed in the previous chapter, the relative scale of life is inversely proportional to the strength of gravity. Rearranging the formula $S_r = 1/g_r$ to predict gravity gives $g_r = 1/S_r$. If we conclude that the scale of life was 2 times as large as expected then gravity would have been ½ = 0.5g (where 'g' equals present gravity) to achieve this relative scale difference. If the scale of life was 1.5 times as large as expected then gravity would be $1/1.5$ = 0.67g - and so on.

In order to compare the relative scale of life in the past with present-day life it is vital to ensure that both the age of any

Chapter 2 - The Giants of the Past

remains found is correct and that a like-to-like comparison is made to minimise any error - even a variation in an animal's activity level can change its optimum size. This type of problem in determining the optimum size means that any value of gravity can only give an estimation of how life's relative scale, and hence gravity, has changed with time.

Fortunately for our knowledge of ancient life the world has supported a vast range of land-based life for hundreds of millions of years and at certain times when these animals and plants died they were naturally buried under mud or sand. This enabled their hard parts to be slowly replaced by rock as their soft parts decayed - a fossil had formed. The process preserved some of them for hundreds of millions of years, so these isolated islands of fossils remained intact as mountains rose and were worn away, and as seas came and went.

The fossil record is very imperfect. Usually there might be detailed fossils of several animals or plants from one brief time in the distant past. Then for tens of millions of years the fossil record is completely blank and nobody can say if these animals became extinct or continued to exist. This is shown by the discovery of animals alive today which have been presumed to have been extinct for millions of years. One such living fossil is the coelacanth as shown in Figure 2.1. The first live example of this fish was discovered off the coast of Africa in 1938, but before this discovery this fish was only known from Cretaceous fossils of over 100 million years ago.

Despite these imperfections in the fossil record, palaeontologists have been able to reconstruct the now extinct range of life in remarkable detail. Although the flesh and muscles of the animals are very rarely fossilised, by studying the shape of the bones the size of the muscles and ligaments can be estimated. Their fossilised footprints are examined and compared with today's animals to estimate their speeds. Their teeth show whether they ate meat or plants, and the fossilised remains of their stomachs tend to confirm this evidence. A piece of fossilised skin can be extended to cover the whole body.

Fig. 2.1 'Living Fossil' Coelacanth

The Coelacanth was thought to have become extinct over 100 million years ago until a 'living fossil' was found in 1938.

Combining all this evidence the paleontologist can propose a very good impression of animals and plants which have been extinct for hundreds of millions of years.

Many of the reconstructions of ancient life are remarkable for the level of detail unveiled about life hundreds of millions of years ago. Take for example the several remarkable fossils of giant land scorpions which have come from a quarry at East Kirton near Edinburgh in Scotland. Andrew Jeram described in the journal *New Scientist* how 400 million to 300 million years ago a number of layers of thin bands of chalk and fine-grained quartz were being laid down in a shallow inland lake.[1] Several scorpions were occasionally washed into the lake from the slopes of nearby volcanoes and these became fossils. The lakes contained little life, perhaps because the volcanic action formed hot springs, so the plants and animals washed into the lagoon soon died and then lay undisturbed for millions of years to enable them to form almost perfect fossils. The scorpions in particular formed good fossils because of a layer of cuticle only 12 thousands of a millimetre thick, which is particularly resistant to decay.

Other evidence comes from the ancient forests that formed the coal of the world. The plants in these coal fields have been crushed and compressed beyond recognition but within the coal are small areas known as coal balls. Once these coal balls are sectioned and etched they reveal the very cells of the ancient plants. The large cells are perfect and the structure is still intact so that the vascular bundles and the support cells can be clearly seen. The structure of the plant is known in such detail that individual plants such as horsetails may be reconstructed. Coal balls have preserved the structure of the plant in microscopic detail.

By the time the dinosaurs had evolved in Cretaceous times, probably about 138 million years ago, a layer of limestone was being deposited in what is now the Sierra de Monte in Spain. Various insects, amphibians, crustaceans, land plants, and on rare occasions birds, were fossilised within this layer of sedimentary limestone. In the journal *Nature* the paleontologist Paul Selden described how he examined fossil spiders under a microscope and discovered that the spiders had been preserved in such fine detail that the bristles on the hairs of the spiders' legs could be seen.[2] He was able to observe the special claws

---

[1] *Jeram, 1990.*

[2] *Selden, 1989.*

Chapter 2 - The Giants of the Past

that are only used to handle the thread to weave it into webs to catch their prey. The detail of the spiders' claws provides good evidence that spiders were weaving webs well over 138 million years ago.

The ability of this limestone to fossilise such fine details has also allowed the appearance of the first birds to be examined. About 140 million years ago the first known bird, *Archaeopteryx,* lived with feathers over most of its body and with only the legs, head and upper neck bare. It had claws on its wings as well as its feet. Its mouth was full of weight-laden teeth and its bones were solid so it must have been heavy for its size.

Some fossil bones of dinosaurs have been preserved in remarkable detail. Scientists in the New Mexico Museum of Natural History have taken sections from the fossil bones and examined them microscopically. There are grains of the mineral-filling silica, and present within this rock are the actual bone cells of the extinct animal. They are the very substance of *Tyrannosaurus rex* which died out 65 million years ago.

Giant Plants of the Palaeozoic Era

The ability of prehistoric land-based life to reach gigantic proportions compared with present-day life started during the Palaeozoic Era, or age of ancient life, long before the dinosaurs had arisen. Some 500 million years ago, while the waters teemed with life, the continents of the Earth were completely barren without any indication of plants or animals. By 400 million years ago the first simple plants pioneered life on land, and within 50 million years these plants had developed into giants which sometimes reached 30 or 40 metres (98 or 131 ft) tall, with trunks two metres (6 ft) in diameter. They formed the first great swamps that produced the coal layers. These primitive plants were giants compared with their modern forms - the horsetails, ferns and club mosses. Present-day horsetails only grow to about 3 metres (9 ft) in height while club mosses seldom exceed 30 cm (12 in) despite the great sizes obtained by their ancestors.

The animals which inhabited these ancient forests were also giants. Among the insects were dragonflies with a wingspan as long as a human arm at over half a metre (18 in), giant mayflies and cockroaches as large as a human hand with the mayflies' wing measuring almost 12 cm (5 in) across, and the length of the cockroaches approaching 10 cm (4 in). Amongst the

Dinosaurs and the Expanding Earth

Fig. 2.2 Ancient Giant Horsetails

Natural sandstone casts of giant ancient plants have a small museum built around them in Victoria Park, Glasgow.

invertebrates were primitive forms of scorpions, millipedes, and spiders. The millipedes in particular grew to gigantic proportions - one species was as long as a cow at over two metres (6 ft) long.

About 330 million years ago, the land where the Scottish city of Glasgow would eventually be built held a swamp where many ancient plants grew. At certain times great quantities of sand and mud swept into the swamp burying the plants. The outside of the plants was hard, but the inside was soft and soon decayed, and as it decayed the sand settled inside the plant to form a perfect mould of the ancient plants. Over hundreds of millions of years the sand turned to sandstone so when Victoria Park in Glasgow was dug up in 1887, the workmen were astonished to find perfect sandstone moulds of these ancient plants.

These remains, as shown in Figure 2.2, give such a realistic impression of the ancient swamps that a small museum was built to display them in Victoria Park. The sandstone casts of these ancient plants are ancestors of the 3 metre (9 ft) tall living horsetail plants except that these ancient plants grew to 15 metres (45 ft) high.

Giant Invertebrates of the Palaeozoic Era

The modern-day dragonfly has changed little, except in scale, in the 300 million years or more since its ancestors flew. These insects are characterised by four large, many-veined, flat wings held horizontally which are structured like a composite honeycomb, giving them strength and flexibility with low weight. They are strong and agile with the largest of the present-day

Chapter 2 - The Giants of the Past

Fig. 2.3 Giant Dragonfly
This life-size reconstruction of the ancient giant dragonfly Meganeuropsis permiana has a 72 cm (28 in) wingspan.
© Werner Kraus 2003

dragonflies reaching a wingspan of up to 16 cm (6 inches). Its young live in water until they mature into adults.

The ancient dragonflies of 300 million years ago were much larger than our present-day species, possibly up to 76 cm (2.5 feet) across the wingspan. The life-size reconstruction of the giant dragonfly *Meganeuropsis permiana,* shown in Figure 2.3, was reconstructed by Werner Kraus for the University Museum of Clausthal-Zellerfeld. He can be seen in this photograph applying the final touches to his reconstruction and clearly demonstrates the impressive scale of this ancient giant dragonfly.

The millipede has existed for at least 300 million years. The present-day species number over 8,000 and they live on any decaying plant matter. Some also eat living plants, and a few are predators. They have as many as 200 pairs of legs and the length varies between 2 mm (0.2 in) long up to 280 mm (11 in) long.

Millipedes leave very characteristic tracks, and these tracks have been found in rocks over 300 million years old. On the shore of an island in western Scotland, there are tracks in the sandstone rocks which match those of the millipede almost exactly. But these tracks are much larger than any present-day species would leave so the size of these millipedes has been estimated at about 2 metres (6 feet) long.

One of the largest known invertebrates discovered so far is *Arthropleura,* as shown in Figure 2.4, which lived in the Upper Carboniferous 340 to 280 million years ago. Werner Kraus has reconstructed this life-size model for a large Diorama display which will include a lot of other creatures of the Carboniferous, including a reconstruction of the giant dragonfly

Fig. 2.4 Giant Millipede
This life-size reconstruction of an ancient giant relative of the centipede and millipede is Arthropleura armata. It lived in the Upper Carboniferous and is the largest known land invertebrate.
© Werner Kraus 2010

Meganeuropsis. These are planned to be displayed by 2012 in Quadrat Bottrop Museum in Germany.[1]

Scorpions were also large during the Palaeozoic Era. Until recently it was thought that they must have lived in water to have become so large but newly discovered fossils have shown how the scorpions evolved lungs instead of gills about 340 million years ago. These finds have confirmed that most Carboniferous scorpions lived on land. Perhaps other fossils will show that scorpions have lived on land for even longer than this.

The biggest complete scorpions so far found were between 35 - 40 cm (14 - 16 in) long. These were nearly 2.5 times the size of the largest species of scorpion alive today. Other fossil fragments indicate that some scorpions may have been up to 70 cm (27 in) long - 4 times larger than today's giant scorpions. The sequences of sedimentary rocks show that scorpions reduced in scale through the Carboniferous until by 300 million years ago they were usually only 30 cm (12 in) long.

The reduction in scale of the scorpions forced changes in their sensors. The compound eyes at the sides of their heads gradually contained fewer and fewer lens. By upper Carboniferous times about 300 million years ago various families of scorpions had between 20 and 40 lens in each eye. Modern scorpions have reduced this to between two and five separate lens. To overcome some of the reduction in eyesight, modern scorpions have developed long slender hairs to sense any slight movements of air.

---

[1] *Quadrat Bottrop Museum für Ur- und Ortsgeschichte Im Stadtgarten 20, 46236 Bottrop, NRW, Germany.*

Chapter 2 - The Giants of the Past

Giant Amphibians of the Palaeozoic Era

The dinosaurs and mammals would not exist until hundreds of millions of years in the future, but their remote ancestors thrived in this gigantic world. They were carnivorous amphibians up to 4 metres (13 ft) long with backbones, four legs, wet skins and jaws that held rows of sharp cone-shaped teeth. These animals resemble present-day salamanders and newts although these modern amphibians are much smaller. The largest living salamander grows to a length of about a metre and a half (5 ft) in the rivers of Japan. A much more common size of amphibian is the 10 cm (4 in) long newt.

These early amphibians of 350 million years ago were limited in size by their ungainly sprawling pose and their wet skins. During the next 150 million years they began to lay their eggs on land and develop dry skins. Most important of all, they began to move their legs under their bodies as shown in Figure 2.5.

As the animals developed more erect postures they grew larger since they were able to support greater body weights. Finally their legs moved right under the body - the dinosaurs, both two- and four-legged, had arrived.

End of the Palaeozoic Era

The evolution of the amphibians into the well-known dinosaurs did not occur without incident. In late Permian time, about 250 million years ago, one of the most devastating mass extinctions of all time occurred. Land life was drastically reduced and about

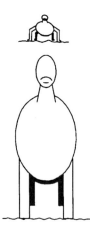

Fig. 2.5 Amphibians to Dinosaurs

Small animals like amphibians are able to use sprawling postures because of their high strength-to-weight ratios. Larger animals like dinosaurs have legs directly beneath their bodies to allow for the reduction in strength-to-weight ratio with increased scale.

half of all marine families disappeared. This mass extinction was used to mark the end of the Palaeozoic Era, although the mass extinction probably occupied millions of years and occurred in several waves.

There have been many great extinctions in the past. While life has existed on our planet millions of species have appeared and disappeared but at times many more species than normal disappear.

Mass extinctions are now accepted as real events. At each great crisis whole groups of organisms have been swept away allowing new species to eventually take their places. These crises have been so great that geologists divide the history of life on Earth into divisions largely based on the major changes in the forms of life.

One characteristic of most mass extinctions is that life evolves rapidly after the period of crisis. This change is often so pronounced that a whole new range of life also seems to appear and predominate as relative peace returns. The next Era of life after the Palaeozoic is dominated by the dinosaurs.

Age of the Dinosaurs

By the time of the dinosaurs the relative scale of life had decreased slightly but it was still high enough to allow the dinosaurs to reach massive sizes. The dinosaurs lived during the Mesozoic Era, or age of middle life, which started over 200 million years ago and lasted for a further 150 million years.

In his book, *Dynamics of Dinosaurs and Other Extinct Giants*, Robert McNeill Alexander, a Professor of Zoology at the University of Leeds in the UK, describes how the study of the mechanics of animals can help to predict how the dinosaurs lived.[1] He notes that it is puzzling how the largest dinosaurs and pterosaurs grew to dimensions larger than present-day life since they were likely to damage themselves due to the disproportionate increase in their weight. The book examines the dinosaurs' stance, gait and speed. These judgements have profound implications on how animals existed.

Some of the sauropods could be amazingly large with long necks and tails to match their size. The largest known complete skeleton of a sauropod is *Brachiosaurus* from the upper Jurassic of about 200 million years ago. It was found in North

---
[1] *Alexander, 1989.*

Chapter 2 - The Giants of the Past

Fig. 2.6 Berlin Brachiosaurus
Brachiosaurus brancai is on display in the Berlin Natural History Museum in Germany. The Berlin Brachiosaurus still remains the largest complete skeleton of a dinosaur ever found. The author can be seen at the front and illustrates the enormous size of these fabulous creatures.

Africa and now stands in the Berlin Museum of Natural History as shown in Figure 2.6. This skeleton is so large that if I was allowed to walk under it I don't think I would be able to reach up to its ribs.

During the 1970s the paleontologist Jim Jensen uncovered two huge new sauropods that he named *Supersaurus* and *Ultrasaurus*. These dinosaurs were named from only a few bones and further study has suggested that these may be huge relatives of the well-known dinosaurs *Diplodocus* and *Brachiosaurus*. Another dinosaur, named *Seismosaurus* in 1991 by the paleontologist David Gillette based on partial skeleton remains, has now also been reclassified as another species of *Diplodocus*. The exact size of these animals is uncertain although they do appear larger than more complete examples, as indicated in figure 2.7. With more complete skeletons like *Brachiosaurus* it is possible to make several calculations regarding the stress levels in various parts of the body.

Dinosaurs and the Expanding Earth

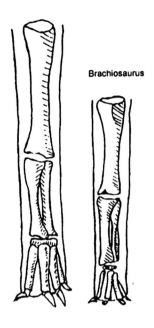

Fig. 2.7 Brachiosaurus and Other Larger Leg Bones

The bones of some dinosaurs appear to be even larger than Brachiosaurus, although the remains are incomplete. Brachiosaurus is still the largest complete sauropod ever found.

The easiest method to use is to compare *Brachiosaurus* with a dynamically similar animal - one which moves and acts in a similar way. *Brachiosaurus* has long front legs and a neck which it seems to have held straight. Its teeth were shaped for eating plants, and fossil twigs have been found where the stomach must have been, so it is most probable that it used its height to eat the leaves of tall trees. *Brachiosaurus* therefore resembles the modern giraffe in form. *Brachiosaurus* was 13 metres (43 feet) tall, which is 2.4 times the height of a full-grown giraffe at 5.5 metres (18 feet).

By calculating the forces acting on various areas of the leg bones, the stresses in the leg bones of *Brachiosaurus* can be compared with the maximum stresses known in present-day large animals. These stress levels vary according to the activity of the animal since the stress levels imposed by walking are about twice those imposed when standing still, while the stress levels imposed by running are about 3.5 times those standing still. The exact sizes of these forces depend on how the animal moves, but using these values it is possible to calculate that the force of gravity must have been about one-third the present value to allow *Brachiosaurus* to run.

The calculations of the stress levels in dinosaur bones illustrate the problems of the dinosaurs' large size. When the

Chapter 2 - The Giants of the Past

Fig. 2.8 Diplodocus's Neck Ligament
In life V-shaped neck vertebra probably held the neck ligament used to keep Diplodocus's neck erect and this enables the ligament's size to be estimated.

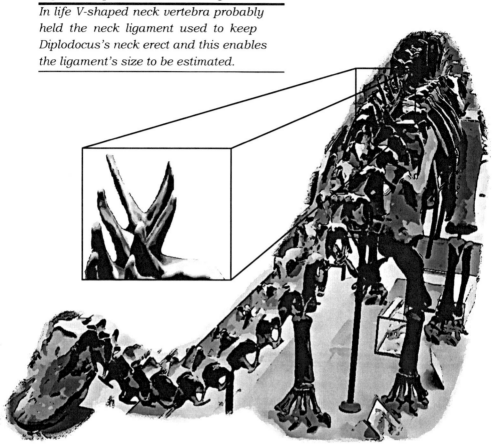

stress levels in the leg bones are calculated for *Brachiosaurus* in the present gravity it leads to the conclusion that *Brachiosaurus* could not have moved faster than walking pace. This is because at any speed greater than walking pace the stresses in the leg would become too high and the bone would fracture and break. Judging from how modern animals move, the maximum speed which *Brachiosaurus* could have run would have been about 50 km/h (31 mph) but the strength of bone would only allow them to move at a maximum speed of 20 km/h (12 mph) in the Earth's present gravity. But *Brachiosaurus* could be just as agile as the large animals of today if the Earth's gravity was less at the time of the *Brachiosaurus*.

Chapter 2 - The Giants of the Past

Fig. 2.9 Stegosaurus
One of the most famous dinosaurs, Stegosaurus, is striking for its large elephantine size combined with large plates and bizarre tail spikes.

metre (3 ft) long that held rows of large sharp teeth - these clearly show that it fed on meat.

There was a variety of vegetarian dinosaurs which all tended to gigantic proportions. There was the bipedal *Comptosaurus* which grew to 6 metres (19 ft) in length. It had strong hind limbs with smaller front limbs, so it may have used all four limbs to browse on vegetation. The bipedal *Iguanodon* was 10 metres (33 ft) long with a peculiarly spiked thumb and was the first dinosaur to be described. *Stegosaurus* was an elephantine-sized four-footed dinosaur with massive triangular plates on its back and spikes on its tail as shown in Figure 2.9. It may have reverted to a four-footed position from its bipedal ancestor but still appears to have been highly active despite being over 6 metres (19 ft) in length.

Mesozoic Era Plants

The earliest plants of the Mesozoic Era, during the Triassic Period, were characterised by gymnosperm plants which included conifers, cycads, and gingko - groups which survive today. There was also still a variety of scale trees (lycopods), from which modern club mosses have evolved, although they were on the decline. The most diverse group of plants was the ferns which formed the undergrowth beneath the larger groups of trees. There were no grasses at this time.

By 220 million years ago the land that is now part of the Amazon held great forests of trees that formed fossilised stone casts of the trees after they were carried downstream to become buried in sand and gravel. These solid fossil trees have been broken into short segments because of the brittle nature of the

Dinosaurs' muscles and flesh have hardly ever been fossilised but it has been suggested by Alexander that the forces in the neck ligament of two other sauropods, *Diplodocus* from the upper Jurassic of about 200 million years ago, and *Apatosaurus*, can be calculated from the shape of the bones which held the ligament.[1] Using the calculations of Alexander, in the Earth's present gravity the neck ligament would only provide about one third of the required force, and the weight of the neck would be nearly enough to break the ligament. These problems are resolved with a reduced gravity.

The neck bones of these dinosaurs have V-shaped neural spines, as shown in Figure 2.8, which would have contained the ligament helping to raise and lower their heads. By calculating the force that this neck ligament was likely to exert and by estimating the volume of the neck and head it is possible to calculate the force of gravity when these animals were alive. For *Diplodocus* it gives a value of gravity of 0.3 the present gravity at 200 million years ago, which allowed it to become 3 times larger than it could in today's gravity.

A modern reconstruction of the sauropods envisages them as relatively agile animals despite the low strengths of their bones. This view of the athletic ability of the dinosaurs was expressed very clearly in Bakker's book *The Dinosaur Heresies*.[2] He believes it is likely that they used their long necks to reach the top of trees and some may have reared up on their hind legs. Their fossil footprints show that their long tails hardly ever touched the ground so they must have held their tail up and this would have helped to counterbalance their long necks. Their fossil footprints also suggest that they travelled in herds with the young in the middle of the herd to protect them from attack by the meat-eating dinosaurs. Studies of the ratio of meat-eating to vegetarian dinosaurs seem to point to a similar ratio to that of modern mammals rather than that of reptiles. This suggests that the dinosaurs were warm-blooded and fast moving like mammals, and not slow like reptiles.

Living with the sauropods were many other species of dinosaurs including large predators that stood on their hind legs. The two-legged carnivorous dinosaurs alive then showed a range of sizes up to the massive *Allosaurus*. Fossil trails show that *Allosaurus* walked on its hind legs by using its massive tail for balance. It was over 10 metres (33 ft) long with a skull a

---

[1] *Alexander, 1989.*

[2] *Bakker, 1986.*

fossil rock but the form of the trees has been so perfectly preserved that it is possible to see the wood grain and the knot holes where branches once grew. Many of these trees were cypress which are related to trees which grow today in the swamps of Florida and Louisiana. The age of the fossil trees when they died can still be calculated by counting the growth rings.

Mesozoic Era Pterosaurs

In the skies flew the largest known airborne animals the world has ever known - the pterosaurs as illustrated in Figure 2.10. One species had an enormous wingspan of over 10 metres (33 ft), as large as a small aircraft, and most tended towards the large size.

The pterosaurs are remarkably similar to the present-day bats in form, although they are classified as two independent groups and are not related to each other. The pterosaurs had similar legs and pelvises and used a stretched piece of skin as a wing. Recently discovered fossils were even covered in fur. Reconstruction of their flight muscles indicates they would be too weak to allow them to fly today.

The early pterosaurs were small with long tails. Many of the later ones had no tail but were much larger than their ancestors. One common large pterosaur was *Pteranodon* which had a wing span of 6 metres (19 ft). Judging from the attachment for its wing muscles it is likely that it was a glider. Perhaps the most unbelievable pterosaur was *Quetzalcoatlus* which was found in the Badlands of Texas in 1971.[1] The fossils indicate an animal

---

[1] *Lawson, 1975.*

Fig. 2.10 Pterosaur
Some species of pterosaur were as large as a small aircraft.

Chapter 2 - The Giants of the Past

with a wingspan of 10 to 13 metres (33 - 42 ft) which means it was larger than some small private aircraft.

Both the pterosaurs and birds which inhabited these ancient skies conformed to the gigantic trend set by the other life forms. They appear to be too large and heavy to fly in today's skies - their flight muscles were too poorly developed and calculation shows that the bones of the larger pterosaurs would tend to crack if they flew today. The impression that these animals were too large and heavy to fly, by comparison with modern bats and birds, has been noted by paleontologists. Some have argued that they never flew, yet in the last chapter I tried to show how all animals tend to evolve into the optimum form for a specific function. Often it is impossible to improve on the design and designers have used the principles of this 'good design' within their own creations. If this argument is followed to its logical conclusion then for me there is no other option but to believe that these ancient pterosaurs and birds must have flown with as much grace as today's birds and bats.

Disappearance of the Largest Dinosaurs

Over the last 50 million years of the Mesozoic Era the reign of the large four-legged sauropod dinosaurs appears to decline with the very largest dinosaurs reducing in number. Their position was taken by a smaller and lighter range of dinosaurs such as *Triceratops* and *Ankylosaurus* among the better known. Although these dinosaurs were smaller than their ancestors they were still gigantic by present-day proportions.

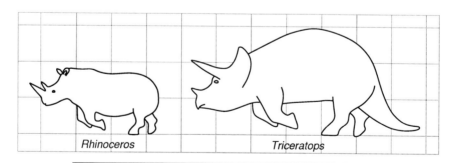

Fig. 2.11 Modern Rhinoceros and Triceratops
A modern rhinoceros and Triceratops both moving and acting in a dynamically similar manner.

Dinosaurs and the Expanding Earth

Triceratops was similar in appearance to a present-day rhinoceros with the most notable difference being that *Triceratops* had three massive horns on its head and it was much larger. *Triceratops* grew to 8 metres (24 ft) in length and its mass was about 9 tonnes, so it was similar to an elephant in mass. Despite its elephantine size, the *Triceratops* seems to be dynamically similar to a present-day white rhinoceros and should have galloped in a similar manner as shown in Figure 2.11. It is difficult to explain how such a large animal was so athletic unless gravity was lower.

Another dinosaur which resembles a modern animal is *Ankylosaurus*. This was an armoured herbivorous dinosaur of the late Cretaceous of North America. It looks like the present-day armadillo, but its superficial resemblance to the extinct giant armadillo, *Glyptodon*, is even more striking. *Ankylosaurus* was only a moderately large dinosaur, being about 5 metres (15 feet) long. The back of its low, flat body was covered by bony plates. At the end of its tail was a thick knob of bone. Some close relatives of *Ankylosaurus* had long, pointed, bony spikes on the end of their tails. Others had bony spikes at the shoulder.

The bipedal carnivorous dinosaur, *Tyrannosaurus rex*, developed a more efficient form by becoming even more like a giant flightless bird. Its forelimbs had reduced in size until they were almost useless while its legs were built for vigorous and prolonged exercise. These leg muscles needed a powerful heart and large lungs to keep them functioning and there is direct evidence that *Tyrannosaurus rex* had a powerful heart and lung system judging from the size of its deep chest.

The duck-billed dinosaurs, *Hadrosaurids*, were a family of large elephant-sized dinosaurs. Modern reconstructions of duck-billed dinosaurs show them as highly active terrestrial animals, as shown in Figure 2.12, in spite of their large size.

Many 'living fossils' of the crocodile and turtle also lived during the Jurassic and Cretaceous and were larger versions of those

Fig. 2.12 Duck-billed Dinosaurs

The elephant-sized duck-billed dinosaurs have remarkably athletic builds for their large size.

alive today. A wide variety of crocodiles ranged in size from small crocodiles to some which grew up to 16 metres (52 feet) in length. But the fact that these animals may have lived totally in water may limit their use for size comparisons. The sea turtles of this time are particularly interesting because they must have come on land to lay their eggs and this fact limits their maximum size. The Cretaceous marine turtle *Archelon* grew to a length of nearly 4 metres (13 feet).

Death of the Dinosaurs

About 65 million years ago the last of these gigantic dinosaurs disappeared from the Earth. Because the fossil record is incomplete there is still argument about whether this extinction happened overnight or over a million years. What is obvious is that the animals it affected were large. The smaller animals appear to have been largely unaffected. The animal families which lived through this period were crocodiles, tortoises, birds, numerous insects, small lizards, small plants and amphibians. These surviving animals were generally smaller than those that became extinct.

In a report in the technical journal *Science* Robert Sloan and his colleagues reported that their studies showed that the dinosaur extinctions in Montana, Alberta and Wyoming were a gradual process that lasted for at least 7 million years.[1] Towards the end of this period the fossils of seven species of dinosaurs and several mammals were found together in a layer which dated from after the supposed sudden extinction at the end of the Cretaceous. Other data from India, the Pyrenees, Peru and New Mexico supports the view that the dinosaurs survived well into the early Paleocene, about 65 to 60 million years ago, in tropical regions.

The fossils of Montana show a progressive reduction of dinosaurs as the mammals increased during the last 300,000 years of the Cretaceous. The demise of the dinosaurs in Montana took place about 40,000 years after the end of the Cretaceous.

During the last 10 million years of the Cretaceous period the dinosaurs of Alberta, Montana and Wyoming progressively reduced in diversity. The ancestors of the new mammals appearing at this time are known to have lived in Mongolia so it

---

[1] *Sloan et al, 1986.*

is generally concluded that the new mammals migrated into the region from there. These new mammals rapidly evolved and then increased in number at one of the most rapid recorded rates of evolution in the fossil record, with the species increasing eight-fold. As the mammal species increased in number during the last 7 million years of the Cretaceous, genus after genus of dinosaur became extinct.

Plants began to reduce in number well before the end of the Cretaceous. They seem to have gradually changed into new forms. At the same time, global temperature decreased and sea levels fell.

The extinction of the dinosaurs was a setback in the flow of life that has mystified many people. A significant fact about this mass extinction is that the animals which became extinct were gigantic and their size has never again been attained by any animal during the millions of years that have followed their reign.

Super-giant Land Mammals

The mass extinction of the dinosaurs ended the Mesozoic Era, and heralded the introduction of the Cainozoic Era, or era of recent life. This is the era when mammals came to dominate the land, and right at the end of the era man emerges.

The tendency for land animals to become gigantic did not end with the dinosaurs. After them came a range of super-giant mammals and several gigantic flightless birds. These immense animals appeared all over the world in different forms but they were undoubtedly giants which approached the scale of the smaller dinosaurs.

In late Eocene times, 38 million years ago, several of these super-giant mammals were present. The largest known was

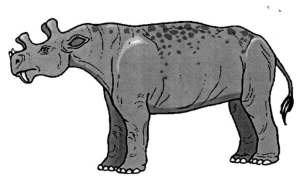

Fig. 2.13 Uintatherium
Giant land animals such as Uintatherium were elephant-sized mammals that seemed more like the rhinoceros in lifestyle.

Chapter 2 - The Giants of the Past

Fig. 2.14 Giant Land Birds
Giant land birds taller than a man, such as Phororhacos (left) and Diatryma (right), hunted the land during the Oligocene.

Baluchitherium, which was a giant 5.3 metres (17 ft) high at the shoulders. Another super-giant mammal, *Indricotherium*, was close to this record, and is believed to be a relative of *Baluchitherium*. Both of these fossil remains were found in Asian deposits.

Amongst the more 'average scale' mammals of that period were many that must still be considered giants. There were the *Titanothere*, a group which approached the elephants in size. *Uintatherium* was the size of an elephant but possessed a horned head full of knobbly bones as illustrated in Figure 2.13. There were 'giant pigs', including *Dinohyus* which was 3 metres (9 ft) long. Amongst the carnivores was *Andrewsarchus*, a large dog-like animal over 5 metres (16 ft) from head to toe.

Several giant flightless birds must be regarded as gigantic when compared with modern forms and among those taller than a man is *Diatryma* from Europe and North America, as illustrated in Figure 2.14. Another type, *Phororhacos*, had an enormous hooked beak over 30 cm (12 in) long and was part of a group called the terror birds for obvious reasons.

Giant Land Mammals and Birds

In more recent times, starting about 2 million years ago during the Pleistocene, numerous giants were present that are recognisable as larger versions of animals alive today. These animals had again reduced in scale but were still giants in our eyes. In Europe and North America were giant bison with horns measuring over 2 metres (6 ft) from horn tip to horn tip. They were hunted by giant wolves known as dire wolves. The giant Irish elk had antlers spanning 4 metres (13 ft) as shown in Figure 2.15; despite its name it roamed over a vast area. There were giant beavers the size of small bears. There were giant

Fig. 2.15 Giant Irish Elk

Giant Irish Elk with antlers spanning up to 4 metres (13 ft) across. Like most of the more recent giants, this seems to be a larger version of today's life.

species of rhinoceros, deer, camel, warthog and ape, as well as the well-known mammoth and sabre-tooth cats. Overhead soared vultures with a wingspan of 4 metres (13 ft).

South America was isolated for much of the last few million years and held a unique set of giants. Charles Darwin was one of the first scientists to comment on the great size of these animals. During his famous voyage on the 'Beagle' he visited South America and observed the remains of:

> ... gigantic land animals... *Megatherium*, the huge dimensions of which are expressed by its name... The great size of the bones of the megatheroid animals, including the *Megatherium*, *Megalonyx*, *Scelidotherium* and *Myldon*, is truly wonderful.[1]

---

[1] Darwin, 1890.

Fig. 2.16 Giant Ground Sloth

South America was home to *Megatherium*, an elephant-sized giant ground sloth.

Chapter 2 - The Giants of the Past

Fig. 2.17 Giant Armadillo
The Giant Armadillo, Glyptodon, was encased in body armour.

The *Megatherium* was a giant ground sloth 6 metres (19 ft) long, as shown in Figure 2.16, and weighing several tonnes. A giant armadillo, *Glyptodon*, was often over 3 metres (9 ft) long. It was encased in body armour and the tip of its armoured tail ended in a mace-like club as shown in Figure 2.17. It looked like a smaller version of the extinct armoured dinosaur *Ankylosaurus*.

At the same time, isolated Australia held several marsupial giants. There were giant kangaroos, a giant marsupial carnivore, and a large platypus. There was a huge marsupial the size and shape of a rhinoceros.

Most of these large animals finally disappeared in the same manner as the dinosaurs of 65 million years ago. Starting 60,000 years ago and lasting for 20,000 years 40% of the large mammals of Africa disappeared. These included the giant baboons and pigs, antlered giraffes, long-horned buffalo, and three-toed horses. The large animals of Europe were next, with 50% of them vanishing, including the mammoth as shown in Figure 2.18, the woolly rhino, the cave bear and the cave lion. These were about 25% larger than their surviving relatives.

The most dramatic extinction occurred in North America, when 70% of the large animals disappeared in a little over 1,000 years. These included mastodons, mammoths, horses, camels, giant sloth, giant beavers, peccaries, dire wolves and others. The major element in the extinction was a clear tendency for the large animals to die out - by far the greatest proportion was those animals with an adult weight of more than 200 kilograms. And those which did survive were smaller than their ancestors.

The Emergence of Mankind

The primates, which are considered by most to be the ancient ancestors of the monkey family, developed during the Cainozoic

Dinosaurs and the Expanding Earth

Fig. 2.18 Mammoth
The Mammoth is clearly recognisable as a large ancestor of the elephant.

Era - from about 65 million years ago to the present. About 50 million years ago lived an animal whose fossil remains show it had fingernails, not claws. It had an opposed thumb on its hands. These are all essential marks of the primates, the family of the monkey, ape and mankind. About 20 million years ago apes lived in East Africa, Europe and Asia. They still lived in trees, despite their size.

By two million years ago, *Australopithecus*, a man-like creature about the size of today's ape, was walking fully erect. *Australopithecus* had a largish brain, and his home was in the Great Rift Valley of Africa, where he lead a short life. The skulls and skeletons show that most of them died before twenty years of age. Alongside the fossil remains of *Australopithecus* are found the rudimentary stone tools which were made by chipping a sharp edge on a pebble. For the next million years this type of tool did not evolve.

One million years ago *Homo erectus* appeared. From their homeland of Africa the ancestors of man spread into new lands. Firstly they had moved into Java by 700,000 years ago, and 400,000 years ago they were in China and Europe. By this time they had developed the use of fire, and the fact that they lived off the land limited their numbers to a low figure, perhaps one million.

Just before *Homo erectus* settled in China and northern Europe the weather became much colder. Great ice sheets formed over much of northern Europe and America. These

Chapter 2 - The Giants of the Past

Fig. 2.19 Sabre-tooth Cat

The large Sabre-tooth Cat hunted many of the giants of its time as well as our ancestors. The last Sabre-tooth Cat only died out about 9,000 years ago.

sheets of ice stayed on the planet for hundreds of thousands of years.

Over 200,000 years ago, during a brief warm period known as an interglacial, the great ice sheets reduced in size as the Earth became as warm as today. These warm conditions did not last and as the Earth gradually cooled the ice slowly returned.

So man entered the most recent Ice Age. At their largest the ice sheets contained so much water that the sea level fell over 100 metres (328 ft), exposing land bridges that had previously been covered by the sea. This enabled man and the animals of the time to walk to other parts of the world, since now Britain was connected to Europe, Europe to North America and Australia to Tasmania.

During this most recent Ice Age man developed spears, harpoons, and at the end of the Ice Age, bows and arrows. Dating from the very end of this Ice Age, about 20,000 years ago, cave paintings found in Spain and southern France show animals that these cave dwellers hunted.

Man would have been in direct competition for food with some of the other large carnivores of the time. The sabre-tooth cat for example, as illustrated in Figure 2.19, roamed a vast area and its fossils are found world-wide. Its most notable feature was two large canine teeth which it used to kill large prey, including the mammoth and great sloth. It was much more robust than today's cats.

It is during the Ice Age that we can picture the typical cave dweller huddled around a fire for warmth. To use this cave they may have had to kill one of the other typical occupants of caves, a cave bear or a cave lion and both of these ferocious animals were larger than their surviving relatives.

By this time we had advanced stone tool manufacture, and there is evidence that the dead were buried with symbolic artefacts, including red ochre, animal bones and flowers. It seems that the range of their speech had developed to produce the 'i', 'u' and 'a' sounds that we know. Reconstructions from the throat of one person who lived about 60,000 years ago show that the size and shape of the larynx had developed into the range of modern man's.

Man's Effect on Wildlife

About 12,000 years ago the Earth entered another geologically brief warm interval, and it is this same interglacial warm period we are presently enjoying. As the ice caps retreated to reveal newly flowering Earth man found that the warm conditions allowed him to develop agriculture and the harvesting of wild grass evolved into the sowing of large ears of wheat. Civilisation had begun.

Some large animals survived the mass extinctions of this period and have inspired many historical references to giants. Notable among these were several giant flightless birds. The largest known was the stout-limbed, *Aepyornis titan*, which stood at 3 metres (9 ft) tall and weighed nearly half a tonne. This flightless bird belonged to a group known as the elephant birds for obvious reasons. Fragmentary remains of bones and eggshell were found in Africa and Europe up to 54 million years old, but while their relatives on the continents became extinct, these large flightless birds appear to have survived on isolated islands. The remains of the *Aepyornis titan* show they were abundant on the island of Madagascar until about 10,000 years ago.

Many of the islands of the world have held a range of bird life which has become extinct soon after people arrived on them. The human settlement of Hawaii, for example, resulted in the loss of half the bird life on the island. The same devastation was caused on New Zealand and the smaller Pacific islands. The islands of the Caribbean and Mediterranean show similar stories. There are another eleven species of birds which have recently become extinct on New Caledonia - almost one third the total population.

Many of the first birds to disappear were large. There are records of species of large pigeons, kagu, megapodes and rails. In New Caledonia the bones were found of an enormous and previously unknown species of game bird, *Sylviornes neocaledonicae*. It was thought to be a distant relative of the emu. Another well-known group, which is now extinct, is the

Chapter 2 - The Giants of the Past

moa species of New Zealand. The largest of these birds approached the elephant birds in size but the last of this species became extinct in the 17th century.

The extinction of many species of animals is generally believed to have been caused by man. When man arrives in any numbers many animals are hunted to extinction. In this way man has greatly affected many of the species of the world. Without man many of the animals which survived the Ice Age would have begun to expand into the new land left by the retreat of the glaciers. This fact makes it unclear how many of the large species would still exist if man had not dominated the Earth.

The Gradual Scale Reduction of Life

According to the theory of evolution we would expect to see the structural strength of all life increasing from the time that life began - and this appears to be correct.

The first plants to evolve had no woody fibres to provide strength. Then they developed a series of cells called tracheids running up their stems which served the dual function of giving added strength with water conduction. With the arrival of the flowering plants these tracheid cells stopped conducting water and started to concentrate on strength.

A similar refinement of body structure occurred within the animal world. The first animals had legs which stuck out sideways. As mentioned previously, this is a very inefficient method of lifting their body weight. It means the animal must remain relatively small so it can easily move its body. As life evolved the legs moved under the animal's body, and because this is a much more efficient lifting method, the animal could become larger. The dinosaurs had reached this stage.

When the mammals finally evolved, they replaced clawed feet with a solid hoof - again a stronger structure.

With all this constant improvement in form for greater strength we might expect that the newer forms of life with improved structures would tend to dwarf the older forms. Yet this is not what happened. The comparison of ancient to modern life shows a general trend of size reduction in relative scale for all life forms over time periods spanning hundreds of millions of years. This can be clearly seen in Figure 2.20 which shows some of the largest animals for different time periods.

Dinosaurs and the Expanding Earth

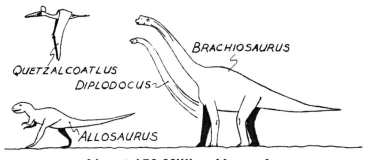

About 150 Million Years Ago

About 70 Million Years Ago

About 40 Million Years Ago

Present Day

Fig. 2.20 Largest Life Over Geological Time
The largest life has been reducing in scale for hundreds of millions of years. This can be explained by an increasing gravity.

Chapter 2 - The Giants of the Past

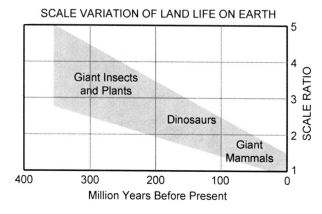

Fig. 2.21 Relative Scale Reduction of Life

The general trend of life's scale reduction can be seen by plotting the relative scale of ancient land life compared to the scale of today's life.

The general trend of scale reduction is noticeable from the observation of the giants of the past, but this trend can be seen more readily by plotting the relative sizes of the giants of ancient times compared to modern life forms. We know that an animal is a dynamic structure which adapts to its environment by attaining an optimum form for its size and this fact can be used so that all animals can be grouped together in similar forms. Once grouped together the relative scale of life can be compared. This method is subject to fluctuations since life has developed with time. Nonetheless, there are some remarkable similarities in ancient and modern forms, such as the fact that the ancient dragonflies and millipedes can still be recognised after 300 million years.

Having established suitable comparisons a graph of the relative scale variation of land life on the Earth can be plotted as shown in Figure 2.21. This graph indicates that when life first crawled onto land over 350 million years ago, it generally obtained a scale of life between 3 - 5 times greater than similar forms of life reach today. By the time of the first dinosaurs this had reduced to between 2 - 3 times and at the end of the dinosaurs' reign it was 1.5 - 2 times the scale of present-day life.

The Changing Force of Gravity

This general scale reduction of all life can be explained if we accept that the gravitational field of the Earth has been increasing over a period of hundreds of millions of years. The low gravitational field on the ancient Earth - a Reduced Gravity Earth - allowed the prehistoric animals of that time to reach a

Dinosaurs and the Expanding Earth

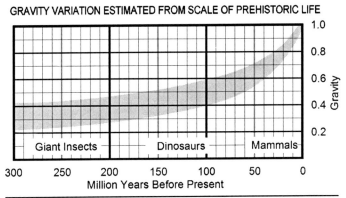

Fig. 2.22 Gravity Increase over Geological Time
An increasing surface gravity on the Earth is indicated by the reducing scale of life over hundreds of millions of years.

much larger size than is now possible. As time passed the Earth's gravity increased and reduced the maximum size which a particular form of life could achieve. As explained in the previous chapter, for a particular form of life the linear scale of land-based life is indirectly proportional to the strength of the gravitational field and this mathematical relationship can be used to estimate ancient gravity from the relative scale of ancient life.

A graph of the Earth's ancient gravity, over periods of hundreds of millions of years, is shown in Figure 2.22 to indicate the probable change in gravity on a Reduced Gravity Earth estimated from the scale of prehistoric life.

Evolution into Smaller Sizes

There is one further interesting thought. Today, most people accept the basic concept of evolution of all life into superior forms. There was, and still is, one major problem with evolution, in that sometimes life appears to evolve in the wrong direction. It actually seems to evolve into inferior *smaller* forms instead of superior *larger* forms.

Darwin and Wallace published a joint paper in 1858 detailing the theory of evolution which is generally accepted today.[1] Later, in 1859, Darwin detailed in his book *The Origin of the Species* his reasons for believing in the evolution of life.[2] In the first

---
[1] *Darwin & Wallace, 1858.*

Chapter 2 - The Giants of the Past

chapters he described how various domestic animals had been modified by humanity. By selective breeding various characteristics had been enlarged or reduced - so sheep, cows, goats and a range of other life, have all been modified by mankind.

Having established that the form of animals can be significantly modified by people, he then showed that there was variation in nature. If any one of these small variations was of use to the animal in the struggle for existence, then nature would select this animal to survive. This was natural selection for the survival of the fittest. If this small variation was then allowed to continue over time then this would eventually lead to new types of animals which were very different from each other. A new species would have formed.

Since the theory of evolution was first proposed, its opponents have constantly referred to the giants of the past to disprove it. In the later editions of his book Darwin asked himself the question of whether the huge monsters of South America could have left behind the smaller descendants of the sloth, armadillo and anteater. He obviously considered smaller to be inferior since he considered the smaller forms to be 'degenerative' and this 'degeneration' of several species seemed to be against all Darwin's theory of evolution.

While Darwin recognised that the giants of South America must be related to the present-day species, he had to reject the idea that the modern forms had evolved directly from the giants and this rejection of life evolving into smaller forms had a profound influence on the basic theory of evolution.

Had Darwin only realised it, the scale reduction of life is a classic example of evolution. Over periods of millions of years life evolved smaller forms to suit the changing force of gravity. Some of these life forms managed to attain this scale reduction using very similar forms, while others were not so lucky and became extinct.

This slow variation in gravity is the explanation of how *all* life reached gigantic scales in the distant past. Although this provides an answer to the mystery of the gigantic scale of these ancient life forms, it raises many more questions, the biggest one being what factor produced this variation in the gravitational field since life first appeared on Earth. This is even more relevant since if the force of gravity has varied in the past, we can expect it to do so in the future.

---

[2] *Darwin, 1859.*

3 - Drifting Continents

The last two chapters have explained why I believe the Earth's gravity has been slowly increasing. But *why* would the force of gravity have changed? Has gravity increased throughout all space or is the increase restricted to the Earth's surface? Has the increase stopped or is it still happening now? These questions can only be answered by establishing the cause of the increasing force of gravity at the Earth's surface.

There are several possible causes of a gravitational increase. At first sight, an easy solution would be if one of the fundamental forces of nature was changing with time. The Universal Constant of Gravity, for example, is present throughout all space and time and is assumed never to vary, so perhaps this fundamental force varies over hundreds of millions of years.

The Universal Constant of Gravity is the fundamental force which tends to make two masses attract each other. The magnitude of the gravitational attraction force between two bodies is directly proportional to the Universal Constant of Gravity and the mass of the two bodies, and inversely proportional to the square of the distance between them. So a change in the Universal Constant of Gravity would change the Earth's surface gravity.

This Universal Constant of Gravity has always been assumed to be the same throughout all space and time and the evidence strongly supports this view. If the Universal Constant of Gravity was changing on the Earth then the same effect would be occurring throughout all space and time. But we know that starlight has been travelling for hundreds of millions of years from distant stars throughout the universe, so any change in the stars that are distant in time and space would highlight any change. These ancient stars follow the same sequence of events

Chapter 3 - Drifting Continents

that younger stars do. Hence the Universal Constant of Gravity must have been the same in the distant past.

Similar arguments would seem to apply for all the fundamental forces of nature. None of them seem to vary over time.

If all the fundamental forces of nature have remained the same then the increase in gravity at the Earth's surface must have been caused by some difference in the Earth. What could be so different about the Earth to cause such a marked change in its surface gravity? One possible solution is that the ancient Earth was much smaller in size and mass. The ancient Earth then expanded in size and mass over geological time to reach its present size - an Increasing Mass Expanding Earth.

So how does an Increasing Mass Expanding Earth explain an increasing gravitational field? There is one obvious fact about all the planets which we can observe from the study of our own solar system. The gravity at a planet's surface is proportional to its size. Small planets have low gravity and large planets have greater gravity. This fact applies to small bodies such as our Moon. So when the Apollo space astronauts visited the Moon they walked in a low gravitational field - about one-sixth of the Earth's. Amongst the planets, Mars is smaller than the Earth and has a lower gravity. Venus approaches both the Earth's diameter and gravity, and the giant planets of Jupiter and Saturn have large surface gravity to match their size. There is a definite relationship between the size of a planet and its gravity, so having established that the Earth's surface gravity was less on the ancient Earth, the almost inevitable conclusion is that the ancient Earth was smaller in both size and mass.

If our Earth is to follow the known physical laws, when the dinosaurs roamed the Earth it was about one half the diameter it is now. Since that time, the Earth has expanded to increase its diameter, mass and gravity. But if this is truly the case, the effects of such a massive increase in diameter should be clearly visible on the Earth.

In late 1987, after I had independently concluded that a smaller less massive Earth would explain the dinosaurs' large scale in a reduced gravity, I found that the theory of an Expanding Earth had already been proposed by a number of Earth scientists including professors and doctors of geology and other sciences. The reasons the Expanding Earth theory had been suggested before had nothing to do with dinosaurs or gravity, but had been developed to account for purely geological evidence that all the continents of the ancient Earth were once joined together in one continental shell on a smaller diameter Earth.

The Expanding Earth theory has had a long history. Many people who have independently 'discovered' the Expanding Earth theory seemed to be unaware that other people had similar concepts. The Russian engineer Jean Yarkovsky suggested the Earth may have expanded in his 1888 book.[1] The Italian geologist Roberto Mantovani also independently published his ideas about an Expanding Earth in 1889 and again in 1909, and these were sufficiently well received to arouse the attention of the French Geological Society many years later.[2] But these ideas about an Expanding Earth seem to have been mostly forgotten for a time.

By 1933 the German engineer and geophysicist Ott Christoph Hilgenberg had also discovered the concept of the Expanding Earth and published his book proposing an Expanding Earth to account for Wegener's Continental Drift.[3,4] But once again this information seems to have been almost forgotten for a period of time.

During the following decades, many other people including Josef Keindl, Lazio Egyed, Allan Cox, Richard Doell, D Van Hilten, Karl W. Luckert, Pascual Jordan, Sam Warren Carey, Bruce Heezen, J. Halm, Lester King, Ludwig Brosske, Kirillow, Cyril Barnett, Kenneth Creer, Ralph Groves, Klaus Vogel, V.F. Blinov and others also rediscovered and developed the concept to various extents. The Tasmanian Professor of Geology, Sam Warren Carey, describes in his book *Theories of the Earth and Universe* how he independently discovered the idea of the Expanding Earth in 1956 and only later, through further research, found the idea had been proposed previously.[5] Karl W. Luckert, an independent researcher based in the USA, also explained in his 1999 book *Planet Earth Expanding and the Eocene Tectonic Event* how he published his first essay on the expansion of planet Earth in 1979 completely unaware of the

---

[1] *Yarkovsky, 1888.*

[2] *Mantovani, 1889, 1909.*

[3] *Hilgenberg, 1933.*

[4] *Wegener, 1915.*

[5] *Carey, 1988.*

Chapter 3 - Drifting Continents

fact that there were several individuals who had independently noticed its expansion before him.[1]

The same idea of an Expanding Earth occurred to Earth scientists who were separated by time, distance and language. It was supported by Americans, British and Australian scientists. There have been several articles and books in German, and Kirillow's reconstruction has been discussed by the Russian geologist Neuman.

These Earth scientists regarded the outlines of the continental blocks as pieces in a gigantic jigsaw which fragmented into numerous bits on the ancient Earth and can therefore be reassembled into its original shape. Using this method, and working independently of each other, Hilgenberg, Brosske and Kirillow attempted to reconstruct the original Earth crust before the last stage of its expansion when the deep ocean floors began to form.

Other Earth scientists such as Egyed, van Hilten, Cox and Doell used the ancient record of the Earth's magnetic field held within certain rocks - palaeomagnetic data as it is known - to calculate the size of the Earth in the distant past.

Various books and articles were written about the Expanding Earth with most authors trying to quantify the amount of Earth expansion. Most authors used the Expanding Earth theory to explain the undeniable fact that an ancient super-continent had broken apart into separate continents that then seemingly 'drifted' into their current positions on the Earth. The 'drift' of the continents was a direct result of the expansion of the Earth.

This separation of the super-continent on an Expanding Earth also explained why the world's ocean floors were all younger than 200 million years. As the Earth expanded in size the new ocean floor filled in the gaps between the continents to create a geologically young ocean floor.

Thus this idea of the Expanding Earth has an interesting history and a small group of dedicated scientists still supports the idea to this day. But this view of an Expanding Earth has not been accepted by most Earth scientists. In fact, today it is one of the most controversial geological theories. To understand why this is so, we need to track the development of knowledge about the Earth.

---

[1] *Luckert, 1999.*

The Foundation of Geology

The view of an unchanging Earth seems to be an ancient idea which has become firmly fixed in mankind's minds, since we have only relatively recently begun to understand that the Earth was a very different place in the past. It was only about 200 years ago, in the eighteenth century, that the basis of all modern geology was first outlined. One founder was the Scottish amateur geologist James Hutton who spent a lifetime looking at the Earth. During his work he concluded that the Earth must be much older than the 6,000 years that some people had inferred from the Bible. After many years of study of the Earth he published his book *Theory of the Earth* in 1795, just two years before he died.[1]

In his study of the Earth he noted that some rocks had been laid down in the distinct sedimentary layers of an ancient ocean floor of sand and mud. These horizontal layers of sand and mud then became compacted and hardened into sandstone and shale as more layers buried them. At some places he saw evidence that these horizontal layers of rock had then been buckled and twisted as the Earth's surface moved. The once flat layers were crumpled by horizontal forces into steep folds in the Earth's crust so that some of these folds were forced above the level of the sea. The process of weathering began on these upturned rocks and over an immense period of time the top surface was eroded away by the action of air and water.

Then this new horizontal surface was again submerged under the sea and new horizontal layers of sand and mud were laid across it. This is now known as an angular unconformity. The evidence for this series of events proved to Hutton that the Earth was unimaginably old, for enormous periods of time must have passed for the Earth to be weathered away in such a manner.

This idea of an immensely old Earth was the revolution that laid the foundation for the science of geology. It enabled mankind to imagine that the Earth was constantly changing over a huge length of time. While the tops of the continents were being eroded by the action of air and water, new sedimentary layers were being built up on the bottom of the sea. These processes occurred for vast periods of time. Land levels repeatedly sank below the surface of the sea and then rose above it again. The various layers and rocks of the Earth were formed by a slow cycle of natural events that occurred over

---
[1] *Hutton, 1795.*

Chapter 3 - Drifting Continents

millions of years. To Hutton it meant that the Earth must be immeasurably old:

> The result is that we find no vestige of a beginning - no prospect of an end.[1]

Today rocks can be dated by their radioactive decay and the oldest rocks known can be dated at over 4,000 million years old.

As these new ideas about an ancient ever-changing Earth were accepted, they influenced the understanding of the fossils of the bones and shells that were sometimes found within these rocks, and visions of the animals that lived in these ancient times emerged. It was obvious that the rocks which held fossils of sea creatures were once under some ancient sea. Then, as the land was thrust upwards the fossil remains became those of land creatures and plants. Sometimes this was seen to repeat itself again and again, so it was clear that such places as the Alps, which held fossil sea creatures, had been thrust upwards enormous distances from their original position on an ancient sea floor.

The Beginning of a New Theory

The English philosopher and scientist Francis Bacon noted the matching shape of the coastline of Africa and South America in his 1620 book, *Novum Organum*, and reflected that:

> ... there are similar isthmuses and similar promontories, which can hardly be by accident.[2]

Much later, the first observations of the geological similarities of the continents were made by the French geographer Antonio Snider-Pellegrini in 1858.[3] He described how the fitting together of the continents bordering the Atlantic explained the occurrence of identical fossil plants in the coal deposits of both Europe and North America. The idea was considered outrageous, mostly ignored, and was soon forgotten.

Other ideas that would explain why the continents fitted together were also mostly disregarded. The Russian engineer Jean Yarkovsky and the Italian geologist Roberto Mantovani independently suggested towards the end of the 1800s that the

---

[1] *Hutton, 1795.*
[2] *Bacon, 1620.*
[3] *Snider-Pellegrini, 1858.*

Dinosaurs and the Expanding Earth

Earth had expanded causing the continents to separate over geological time.[1]

By the early 1900s the most widely held geological ideas about the Earth had begun to resemble those of today, but in some ways they were very different. The Earth was believed to be a solid sphere in which the positions of the continents had remained fixed. There were land bridges that rose up out of the sea to connect continents together and then disappeared back into the sea. It was a theory that was defended by the most respected scientists of the day, even though we now know it was completely wrong.

In 1910, the American geologist Frank B. Taylor proposed that the land masses of the northern hemisphere had moved bodily away from the poles towards the equator.[2] This movement had caused North America to split away from Greenland. Apart from these two land masses, he made no attempt to reassemble the jigsaw puzzle of the continents.

In 1911, Howard B. Baker postulated a single continent and illustrated this with the 'displacement globe' as illustrated in Figure 3.1. In 1912 he summarised his thoughts in more detail in his book, *The Origin of Continental Forms*.[3]

Baker suggested a single continent which split to form the continents of today. The northern continents are fitted tightly

---

[1] *Yarkovsky, 1888: Mantovani, 1889, 1909.*

[2] *Taylor, 1910.*

[3] *Baker, 1912.*

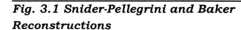

Snider (1858)

Fig. 3.1 Snider-Pellegrini and Baker Reconstructions

Various authors began to reconstruct the continents as one gigantic super-continent on a constant diameter Earth around the turn of the 1900s.

Baker (1911)

Chapter 3 - Drifting Continents

together and the Mediterranean basin is closed up with the rotation of Italy, Spain and Newfoundland, thereby bringing North America next to North-west Africa and close to South America. There are good indications that all these land masses were once connected in this way.

Completely independently of this previous work, in 1912 the German meteorologist Alfred Wegener proposed in public that the continents had drifted apart. He amassed evidence that the continents had once been joined in one gigantic land mass 300 million years ago. This super-continent, which he called Pangaea, then broke up into the smaller continents of today to finally end up in their present positions.

The spark which gave Wegener his new theory, which he called Continent Drift, was the realisation that the outlines of the continents fit together like an oversize jigsaw puzzle. Wegener appears to be completely unaware of any similar previous observations, but it was Wegener who searched for further evidence which would either prove or disprove the theory of Continental Drift.

The concept of Continental Drift first came to Wegener in the year 1910, while considering the map of the world. At first he considered the idea that the continents of South America and Africa were once joined to be improbable. Then, after learning in 1911 of the similarity of prehistoric animals and plants on these continents, he examined the research in the fields of geology and prehistoric life. The evidence he found convinced him. By the beginning of 1912 he had delivered an address on the idea and had published two papers on the origin of the continents in scientific journals. The First World War then interrupted his work, but in 1915 he published a German only edition of his book, *Die Entsehung der Kontinente und Ozeane*, (The Origin of the Continents and Oceans).[1] The book went through four editions with each edition being thoroughly revised. By the fourth edition of 1929 Wegener was able to quote evidence from other workers in the field and this edition was also translated into English and other languages.[2]

The similarity of ancient fossils found on continents as remote as Africa and South America, Europe and North America, Madagascar and India, had already been explained by the current theory of the time - the land bridge theory. In this theory the continents were connected by bridges of land that spanned

---
[1] *Wegener, 1915.*
[2] *Wegener, 1929.*

Dinosaurs and the Expanding Earth

Fig. 3.2 Continental Blocks 'Float' on the Denser Mantle

The lighter continental blocks appear to float upon the denser material of the ocean floor. Some parts of the continental blocks extend under the ocean to form the continental shelves.

the oceans. This theory assumed that a land mass the size of a continent sank beneath the sea, and so this was one of the first misconceptions that Wegener corrected. He explained how modern science of the time showed that the ocean floor and the continents are fundamentally different. By comparing the average heights of the continents and the ocean floor, he showed there were two mean elevations of land. One level was the abyssal plain, or flattest surface, of the ocean floor, while the average level of the continents was set at 4,800 metres (15,750 ft) above this level. This fact was so clear that Wegener said that:

> ... in the whole of geophysics there is probably hardly another law of such clarity and reliability as this.[1]

If the oceans and the continents were composed of the same material then a graph of the elevations would reach one high point. This is a standard result of the laws of probability. In practice, there are two peaks, one centred on the continent elevations and the other centred on the sea floor. This shows that the continental blocks are composed of a lighter material than the denser ocean floors. From these facts Wegener deduced that the lighter continental blocks, including the continental shelves as shown in Figure 3.2, were effectively floating on the denser material of the ocean floor. Of course, the idea that a whole continent can float on a solid material like rock takes some explaining, even today.

One main reason why the Continental Drift theory was rejected for so long was the difficulty of envisaging these massive continents of the world floating on the denser ocean floor. This ocean floor is made of extremely solid rock, so how could even the smallest of the continents float on the ocean floor? The

---

[1] *Wegener, 1915.*

Chapter 3 - Drifting Continents

answer to the riddle lies in the basic assumption that the Earth's rocks are solid over a long period. In fact, over long time intervals, most materials lose their rigidity and begin to flow in a plastic manner. This property is known as creep in the engineering world, and must be taken into account when designing structures in seemingly solid materials like steel. Over periods as small as 15 to 20 years the maximum stress that can be applied to the material of a steel pressure vessel must be decreased to keep the flow of the material within acceptable limits. Various engineering structures such as pressure vessels, cranes, bridges and aeroplanes are regularly examined for the tell-tale cracks which appear as the metals flow due to the forces acting on them.

The almost instinctive assertion that rock is always solid is completely wrong. It is an illusion that is produced by assuming that because a rock is hard over the short time periods that we observe it, then it is also hard over much longer periods. In practice, rocks tend to flow like a liquid over long periods of time.

One example Wegener used was the fact that all the land of Scandinavia was steadily rising. During the recent Ice Age great glaciers weighed down that part of the continental crust. As the ice melted, the whole of Scandinavia became light enough to begin rising, and it is still rising today. Other examples of how the continents are floating on the crust of the mantle are plentiful. The western coast of Australia has been sinking for the past 400 million years; the Colorado Plateau of the western United States has risen by 2 kilometres (1.2 miles) during the last few million years.

Once Wegener had shown the fundamental difference between the continental blocks and the deep abyssal plain of the ocean floor it becomes impossible to believe that the ocean floor used to be continental land bridges. After he had disproved the land bridge theory, the only other explanation for the similarity of ancient fossils was if the continents were once joined in the one gigantic land mass of Pangaea. The theory of the drifting continents explains why ancient fossils such as earthworms, snails and plants were almost identical on continents which are now far apart. Since these creatures could not have flown or swum across vast oceans the continents must have been connected at some time.

Fossil evidence shows that forms of life that seemed incapable of traversing broad oceans existed on all the southern

Dinosaurs and the Expanding Earth

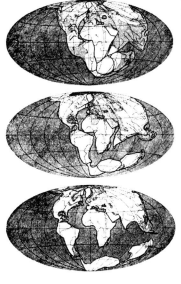

Fig. 3.3 Wegener Reconstructions
Wegener reconstructed the continents as one gigantic super-continent on an ancient Constant Diameter Earth. The light stipple indicates shallow seas.
© *Alfred Wegener 1915.*

continents of Pangaea. One of these was a small reptile, *Mesosaurus*, which inhabited both Brazil and South Africa during Palaeozoic time. The *Glossopteris* flora is even more widespread and occurs on all the southern continents. This tree-like plant has pea-sized seeds that could not be blown across vast oceans. More modern mammal-like reptiles also existed on most of the southern continents at this time.

He found more evidence in the massive valleys left in the Earth by ancient glaciers. When movement of an ancient ice flow is mapped in the Carboniferous period of 300 million years ago it shows that now tropical regions such as India and Africa were under ice. Continental Drift provides us with the theory of how this can occur. The continents must have moved. Figure 3.3 shows Wegener's proposed rearrangement of the continents from 300 million years ago to the present day.

Additional evidence is found in the rocks of the southern continents, since the rock record on each of the southern continents is similar. There are remarkable similarities in the sequences of late Palaeozoic and early Mesozoic rocks on both sides of the Atlantic. Sediments deposited by glaciers are overlain by coal beds which hold the *Glossopteris* fossil flora, these in turn are overlain by desert deposits, and finally at the top of the sequences are dark rocks formed by lava.

Wegener had made many expeditions investigating the drift of continents. To obtain new evidence for the drift of Greenland,

Chapter 3 - Drifting Continents

Wegener planned a tough and dangerous expedition there for 1930. The main expedition discovered many new facts, but in November 1930, the harsh conditions of the Greenland ice cap claimed the life of Alfred Wegener.

Although Wegener had provided a considerable amount of evidence for the drift of continents, he was never able to explain why they had moved such vast distances. He appeared untroubled by this fact as he stated:

> The Newton of drift theory has not yet appeared. His absence need cause no anxiety; the theory is young and still treated with suspicion.[1]

Unfortunately, the absence of any suitable means of explaining the drift of continents meant that the theory was never wholly accepted. The Continental Drift theory had been the subject of heated debate in Germany when Wegener's book was published in 1915. The first editions were only published in German so most of the English-speaking world remained ignorant of his work until an English translation was published over a decade later. The final edition, which was also translated into French, Spanish, Swedish and Russian, started a worldwide debate about Continental Drift.[2]

But then, after Wegener's death in the snow of Greenland, interest declined. It was not until nearly 30 years later, in the late 1950s, that startling new evidence was obtained to prove the theory of Continental Drift beyond any doubt.

Although there was scant interest in Continental Drift several people still continued to propose new ideas that explained the concept of drift. One explanation for Continental Drift was proposed three years after Wegener's death in a 1933 book, *Von wachsenden Erdball*, which roughly translates as 'The expanding globe' or 'The expanding Earth'.[3] Ott Christoph Hilgenberg, a German engineer and geophysicist, gave details of how the continents of the world could have formed from one continuous continental shell on a smaller diameter Earth. Then, as the Earth expanded in size, the continental shell was split into the various continents of today. As the expansion continued the continents drifted further apart until they reached today's positions as shown in Figure 3.4.

---

[1] *Wegener, 1915.*
[2] *Wegener, 1929.*
[3] *Hilgenberg, 1933.*

Dinosaurs and the Expanding Earth

Fig. 3.4 Hilgenberg Expanding Earth Reconstructions

Hilgenberg suggested in 1933 that the continents may have been moved by the Earth's expansion in size over geological time and constructed globes to demonstrate the concept.
© Hilgenberg 1933.

Regrettably, when I searched for Hilgenberg's book in the early 1990s it was so old that the British Library was unable to locate a copy - even in German. Copies from two different locations had disappeared and, unlike Wegener's book, it had not been reprinted later on. The only information I could obtain about this book for the first edition of *Dinosaurs and the Expanding Earth* was a map of how the continents of the world can be reconstructed to fit together on a smaller diameter Earth. But in 2003 a book called *Is the Earth Expanding?* was published in honour of Hilgenberg and this provides valuable background information about Hilgenberg's thoughts and work.[1] It is also an excellent source of additional information about the Expanding Earth.

Although most scientific opinion had swung against Continental Drift, there were still some notable supporters. The South African geologist, Alexander du Toit, published some of the best geological arguments for Continental Drift in his 1937 book *Our Wandering Continents*.[2] He believed that there was no convincing interpretation of Palaeozoic fold systems in the southern hemisphere that did not include the movement of the

---

[1] *Scalera & Jacob, 2003.*

[2] *Du Toit, 1937.*

Chapter 3 - Drifting Continents

continents, and whether the causes for such drift were known or not, there was a great and fundamental truth embodied in the revolutionary hypothesis.

Du Toit recognised that the continents of the world could not be reconstructed in one land mass to stretch across a sphere of the same diameter as the present Earth. To overcome this problem he reconstructed two groups of continents into two separate super-continents as shown in Figure 3.5. The northern continents of North America, Greenland, Europe and Asia (except India) were reconstructed as Laurasia, while the southern continents of South America, Africa, Antarctica and Australia were reconstructed as Gondwana. This separation of the land masses into two separate areas allowed the known evidence for the weather and life of these areas to be treated separately.

There was a very good reason why the reconstructions of the southern continents had to include India. The unanimous opinion was that Africa, Madagascar and India had once formed one land mass. Various faults in the Precambrian basement can be seen to extend through Africa, Madagascar and India, and rocks found in these three places are found nowhere else in the world, except Antarctica. These sequences can be matched with each other. Various fossil remains found in the rocks also match.

The removal of India from the rest of Europe and Asia, and indeed the whole separation of the two super-continents, is only necessary if the Earth was the same size as the present. With a smaller diameter Earth all the continents would fit together into

Laurasia

Gondwana

Fig. 3.5 Laurasia and Gondwana Reconstructions

Du Toit was forced to reconstruct two separate super-continents, Laurasia and Gondwana, to allow the ancient North and South weather systems of the past to be centred over the appropriate continents. But on a smaller diameter Earth the weather patterns would be correct.

one land mass while the weather patterns of that time would also fit the new positions. However, the theory of Continental Drift was still not accepted, so the more radical theory of an Expanding Earth never stood a chance.

At the end of his book du Toit examines reasons why the continents could have drifted apart. He looks at a contracting Earth, rotational forces and convection currents. Du Toit was probably not aware of the Italian, Russian or German work on the Expanding Earth, but in a presidential address to the Astronomical Society of South America in 1935, J. Halm explained the evolution of the Earth with his own Expanding Earth model.[1] This prompted du Toit to mention very briefly that:

> Halm has given some attractive reasons, based on stellar evolution, for an Expanding Earth, and has indeed employed that hypothesis to explain the development of the continents from a former Pangaea. While such expansion would very simply account for continental fracture, a difficulty arises through the varying amounts by which the blocks have been separated. Certain of Halm's conclusions are nevertheless striking and worthy of a close study by geophysicists.[2]

Arthur Holmes was a British geologist who also championed the theory of Continental Drift while it was still unfashionable. His book *Principles of Physical Geology*, first published in 1944, ended with a complete chapter on Continental Drift.[3] He suggested that the drift might be caused by giant Convection Cells in the Earth's mantle, a concept he had first suggested in a lecture to the *Geological Society of Glasgow* in 1928. He proposed that radioactive energy in the Earth's core would heat the Earth's mantle forcing hot magma to the Earth's surface. This material then moved across the face of the globe until it had cooled enough to sink down again into the Earth's mantle. The slow movement caused giant Convection Cells to form that could drag the continents across the face of the globe causing Continental Drift. The concept made little impact on geological thinking at the time.

---

[1] *Halm, 1935.*

[2] *Du Toit, 1937.*

[3] *Holmes, 1944.*

Chapter 3 - Drifting Continents

The vast majority of geologists rejected the concept of Continental Drift as 'an absurd daydream' but startling new evidence about the ocean floor soon began to change this view.

Continental Drift is Rediscovered

After the Second World War the ocean floor began to be studied with the use of echo sounding equipment. During the 1950s the data of more than 50 expeditions was incorporated into a map of the ocean floor by a team of geologists based at Columbia University in the USA. This was published in book form in 1959 as *The Floors of the Oceans* by Bruce Heezen, Marie Tharp and Maurice Ewing, to give details which had never been seen before.[1] From these maps it was clear that the ocean floor was divided into three major regions - the continental margin, the ocean basin floor and the mid-oceanic ridge.

The continental margin included the areas above land and some areas which were submerged under the ocean. These continental seas were never very deep and their boundary was marked by the steep edge of the continental block dropping down to the ocean floor. The deepest parts of the ocean basin floor lie next to the continents. These are known as the abyssal plains since they are the flattest surfaces of the Earth. These abyssal plains are apparently built-up deposits washed down from the continental shelves.

Further out towards the centre of the ocean, the ocean floor becomes uncovered to reveal the abyssal hills. These hills rise from the abyssal floor through a series of scarps until they reach the mid-oceanic ridge. This mid-oceanic ridge is the largest feature of the Earth. It consists of a broad fractured arch through the middle of the ocean. The level of the ocean floor rises until it touches the sides of a rift valley which forms a long cleft 20-35 km (12-21 miles) wide and 1-3 km (0.6-1.8 miles) deep. These rift valleys are volcanically active.

The similarity of the mid-Atlantic ridge and African rift valleys prompted Heezen and Ewing to compare the topography and seismic activity of these two areas. They concluded that the two areas were the same structure, and must form parts of the same continuous structural feature. Since the African rift valleys seem to be the result of the stretching of the crust it seemed that the ocean floor was splitting apart. This could be explained if

---

[1] *Heezen, Tharp & Ewing, 1959.*

extensive vulcanism and intrusion of new material along the mid-Atlantic ridge had been constantly filling the newly-formed gaps.

During the late 1950s Heezen described how he believed the southern Atlantic first formed. The super-continent consisting of Africa and South America was first broken up by a rift valley, similar to the one which is presently forming in East Africa. Over millions of years, the continuing separation of the two continents allowed the sea floor to widen as lava formed a new floor at the centre of the ocean ridge. Because of the method of formation the age of the ocean floor gradually became younger towards the centre of the ocean. At the very centre of the ocean ridge the new material was still forming as lava welled up from the depths.

During all the time since the two continents of Africa and South America were joined, the position of the mid-Atlantic ridge still corresponded with the original rift which first formed. Even today the mid-Atlantic ridge still appears to lie along the same line as the edges of the continents.

The geologists of the southern hemisphere, who had seen the evidence for drift with their own eyes, lead the number of scientists who became convinced of the reality of drift. One of these was Sam Warren Carey, who was Professor of Geology at the University of Tasmania. In 1958 Carey demonstrated the fits of Africa and South America on a large, 76 cm diameter globe with continents made from Perspex so that they could be slid over the surface of the globe. Using this simple model he was able to show the almost perfect fit of Africa and South America.

Carey showed that the best fit of the two continents of South America and Africa was obtained if the edge of the continental shelf was used instead of the present coastlines. This fit was so good that there can be no other explanation - these continents were once joined.

In the same year that he demonstrated the fits of the continents on a large globe, Carey organised an international conference on the drift hypothesis. His detailed reconstructions had proved to him that the ancient continents only fit together properly on a smaller diameter Earth, allowing the edges of all the continents to fit neatly together, so he publicly stated that he believed that the drift had been caused by the Earth expanding in size:

Chapter 3 - Drifting Continents

> It might lead ultimately to an absurdity whereupon I would abandon it, but it has not done so yet. On the contrary it prompts me to ask the physicist to seek an Earth model, which will expand at an increasing rate.[1]

But accepting that the Earth had expanded was difficult at that time. The scale of the problem can be seen if we consider that if the Earth has doubled in diameter, its surface area is four times greater and its volume is eight times greater! Where has this extra volume come from? One of the first reasons proposed for an Expanding Earth was that the very force of gravity was weakening over hundreds of millions of years. Other factors examined at the time were that the Earth may have heated up. Or perhaps it changed in volume with chemical changes in the mantle of the Earth.

One idea which some people thought might explain the Expanding Earth was the concept that the Universal Constant of Gravity was weakening over time. This theory had been first proposed in 1937 by Paul Dirac - a British theoretical physicist who made fundamental contributions to the early development of quantum theory.

This idea of a weaker gravity was discussed in detail by Pascual Jordan, the Professor of Physics at the University of Hamburg, in connection with the geophysical implications, including the effects on the long-term climate of the Earth. With a German edition in 1966 and an English edition in 1971 of his book *The Expanding Earth,* Jordan tried to inspire a wider circle of interest in the Expanding Earth.[2] For more than a decade, Jordan had tried to stimulate interest in the idea and only presented his conclusions in book form when it became clear that nobody else would take the trouble. In the introduction he describes how his main objective was to find out if the Expanding Earth hypothesis was right or wrong within his lifetime.

Jordan proposed that the ideas of Dirac would explain why the Earth had expanded. Dirac's hypothesis was that the very Universal Constant of Gravity was growing weaker with time. With a higher gravity the Earth was more highly compressed. As the Universal Constant of Gravity weakened, the Earth was forced to expand. This also explained the subdivision of the Earth's surface into continental shelves and deep-sea basins.

---

[1] *Carey, 1958.*

[2] *Jordan, 1971.*

Jordan again queried the composition of the continental land masses and the ocean floor. Why are continents and ocean basins different? Wegener's observations, and others, had shown the outstanding characteristics of a two layer structure of the Earth's surface. There was a sharp boundary between the high-level regions of the continental blocks and the deep sea floor. The work of Heezen had also shown the typical steep continental slope down to the ocean floor.

Jordan was convinced that the ocean's floor had been created as the Earth expanded and he clearly stated his views:

> In several places the rifts continue to the continents where they become rift-valleys, and this confirms our interpretation of them as signs of the Earth's expansion. There are clear indications that the sides of these rift valleys are gradually separating ... We can consider the existence of the system of rifts and cracks, and its importance as proof of expansion, as conclusively established despite the short time since its discovery. There is no doubt upon closer examination that this is one of the most magnificent phenomena shown by the Earth.[1]

Jordan's proposal that the expansion of the Earth was caused by the Universal Constant of Gravity weakening with time was similar to the cosmological ideas of the British astrophysicists Fred Hoyle and Narlikar.[2] But this idea was not liked by other cosmologists - the higher Universal Constant of Gravity in the past would make the Sun considerably hotter than the present since the luminosity of the Sun is proportional to the seventh power of the Universal Constant of Gravity. This much greater solar heat output would have prevented the evolution of life in the time we know it occurred. If anything, the indications of ice sheets in the past show a Sun which was cooler than the present.

Even as scientists considered how the Expanding Earth might be related to Continental Drift, other new evidence continued to come from the painstaking efforts of scientists to examine the ocean floor. The results of this examination proved that the ocean floor was progressively younger towards the centre of the ocean. In fact, none of the ocean floor is older than about 200

---

[1] Jordan, 1971.
[2] Hoyle, & Narlikar, 1971.

million years. This is far younger than the continental blocks, some of which were formed over 3,800 million years ago.

As well as the echo soundings, the research ships had recorded the total magnetic intensity of the sea floor. These were seen to vary in rapid jumps. The reason remained a mystery, but in 1963 two British scientists, Fred Vine and Drummond Matthews, explained these in a way that enabled the age of the whole of the ocean floor to be easily mapped.

Vine and Matthews proposed that as hot lava spread out from the central mid-ocean ridges it cooled and recorded the Earth's magnetic field at that time. Since the Earth was known to reverse the direction of its magnetic field every several thousand years to a few million years or so, the ocean floor would act as an ancient tape recorder of the Earth's magnetic field as it varied over time.

These ocean floor magnetic recordings were then mapped and combined with geological age dating of rock samples dredged up from the ocean floors to enable detailed geological maps of the age of the Earth's ocean floor to be easily produced. These maps showed that the ocean floor was younger at the ridges and progressively older towards the continents. The ocean floor on either side of a mid-ocean ridge was also a mirror image of itself and this reinforced the evidence for the theory.

Later, in 1965, computer models of the continents by the British geophysicist Edward Bullard showed that the best fit between South America and Africa was obtained by using the 2,000 metre (6,562 ft) depth contour, which is about half way down the continental slopes. By 1969, Antarctica, Australia and India had been included in the reconstruction of the continents. The fit was so good that it became impossible to explain the fit in any other way than Continental Drift. All the continents had once formed one land mass which had subsequently been torn apart.

The Expanding Earth Theory is Rejected

Although the evidence for Continental Drift had become overwhelming by the 1960s, the reason the continents had drifted so far apart still needed an answer. Despite all the scientific efforts proving the reality of Continental Drift, the theory of the Expanding Earth was still not generally accepted. It was too astonishing to be recognised by most Earth scientists. Not many years ago the Earth was thought to have been static

Fig. 3.6 Convection Cell

The currently widely accepted theory of how the continents have moved on a Constant Diameter Earth proposes that giant convection cells form within the Earth with the continents floating on the top.

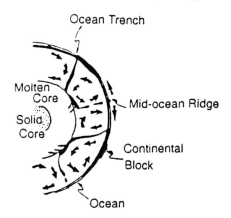

SECTION THROUGH A CONVECTION CELL EARTH

for millions of years and it had only very recently been accepted that the continents were once joined. So despite the best efforts of some of the most respected Earth scientists in support of the theory of the Expanding Earth - notably Carey and Heezen - most scientists favoured a more conservative answer.

This other answer came as a sea floor which continually renewed itself every two hundred million years. Harry Hess, an American geology professor at Princeton University, suggested that Arthur Holmes's widely ignored concept of giant Convection Cells in the mantle of the Earth might provide the answer.[1] He proposed that the new ocean floor spread out from the central ridge towards the continents. Then as it reached the continents it sank down under the continents and was consumed deep within the mantle of the Earth, only to be recycled millions of years later as new ocean floor. It was a theory which was readily accepted as more in keeping with the traditional views of a Constant Diameter Earth. It is this theory which is accepted today.

For over 30 years the reality of Continental Drift had been denied because the ocean floors were considered to be too solid to allow continents to drift over them. Now scientific opinion radically altered so most geologists began to believe that not only the continents, but the whole of the ocean floor were part of giant Convection Cells which spanned the world as shown in Figure 3.6. The whole ocean floor needs to be subducted into the trenches to maintain the concept of a Constant Diameter Earth.

---

[1] *Hess, 1962.*

Chapter 3 - Drifting Continents

Fig. 3.7 Intermittent Subduction Zones

The subduction zones around the Pacific ocean are not continuous. They are only intermittent areas that could not devour whole oceans. The shaded areas show the intermittent subduction zones around the Pacific 'ring of fire'.

Without this subduction of the whole ocean floor the Earth would expand as new ocean floor was created.

When the theory of giant Convection Cells was first proposed the knowledge of the sea floor was a new science, so since then geologists have been looking for new facts to prove or disprove their theories. These new facts have posed some difficult questions.

The most widely accepted mechanism to move a giant Convection Cell over millions of years was the internal heat within the Earth. It was suggested that less dense hot magma would rise at the mid-ocean ridges while more dense cold rock sank down at the continental shelves to move the giant Convection Cell. But Jordan pointed out that a heat Convection Cell of the whole ocean floor would need to overcome the internal friction of all the world's rock. The only energy available for this work comes from the heat at the ocean rifts, so this heat energy must be converted into mechanical work and the process must obey known thermodynamic laws. From the known values he calculated that the efficiency of such a system could never be high enough. The amount of measured heat would never be enough to produce any convection currents in the mantle of the Earth.[1]

A giant Convection Cell would need immense forces to drive the convection over millions of years. The forces acting on such a convection system would certainly be able to buckle and fold large sections of the Earth to produce mountains. It was

---
[1] Jordan, 1971.

suggested that some continental mountains were formed in exactly this way.

However, there are some problems with this proposal. The ocean floor is thinner than the continents so they should be the first to buckle to form mountains. This does not happen. It is within the continental blocks, where the stress would be the lowest, where most mountains form. The fact is that the ocean floor is one of the flattest places on the Earth.

Other scientists also pointed out that the place where the old crust was supposed to be consumed, the so-called 'ring of fire' subduction zone which surrounded the Pacific Ocean, was not continuous. It was more like a small number of discontinuous lines around the Pacific Ocean as illustrated in Figure 3.7. The significance of this is that if the region of active subduction zones outlined by volcanoes and earthquakes does not form a continuous line, then how can the whole floor of the Pacific Ocean be subducted into the mantle? The question remained unanswered, but it did not shake the belief in the giant Convection Cell model.

Another problem was the difference in length between the mid-ocean ridges generating new ocean floor and the subduction zone. The ocean floor being created is about three times longer than all the subduction zones. How can this be? They should be exactly the same length on a Constant Diameter Earth.

Carey had not been idle. He had published various articles and organised conferences on the subject of the Expanding Earth in 1958, 1975, 1976 and 1983. With the last of these Expanding Earth conferences Carey was able to act as editor for several contributors. In 1976 Carey also published his first book on the novel theory - *The Expanding Earth*.[1]

Other supporters gave additional reasons to believe in an Expanding Earth. In 1979 a description of Continental Drift, provided by Hugh G. Owen of the British Museum, gave several reasons why Continental Drift could only be reconciled with an Expanding Earth.[2]

First, the re-fitting of the continents into the land mass present at the time of the Pangaea was precise on a smaller diameter Earth, whereas reconstructing the continents on an Earth of the present diameter obliged Earth scientists to imagine that a wide ocean, known as the Tethys Sea, separated

---

[1] *Carey, 1976.*
[2] *Owen, 1979.*

Chapter 3 - Drifting Continents

the continents of the Eurasian plate from those of the African and Indian plates as shown in Figure 3.8.

The second point was that the distribution of the land-living *Dicynodon* reptiles was exceptional if a wide sea separated greater India from China. These reptiles lived about 200 million years ago in the Triassic period as land-living, herbivorous animals which lived in areas of South Africa, South America, Greater India, Antarctica and China. If a wide sea separated Greater India from China, it is difficult to imagine how these animals could migrate between these two land masses. With the continents reconstructed on a smaller Earth, and the disappearance of the Tethys Sea, the problem no longer exists.

The third point was that although the subduction in the North Pacific Ocean superficially supports the idea of an Earth of constant dimensions, the South Pacific Ocean has vast areas of oceanic crust which can only be explained with an Expanding Earth. From the late Cretaceous to the present day vast areas of new oceanic crust were generated around the southern ocean ringing Antarctica (parts of the Indian and South Atlantic Oceans). Since there are no subduction areas which could have compensated for these new areas of ocean floor, the only remaining explanation is that the Earth has expanded as this new ocean floor formed.

In his book *Atlas of continental displacement - 200 million years to the present,* Owen gave detailed maps which compare

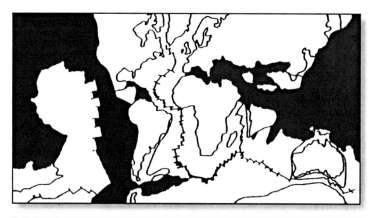

Fig. 3.8 Gaps on an ancient Constant Diameter Earth

Fitting the land and oceans which existed 65 million years ago on an Earth of today's diameter leaves vast gaps in the ocean floor.

results to a Constant Diameter Earth model. The results were what they expected to see and few questioned them.

The Australian geologist James Maxlow discussed this problem in detail in his 2005 book *Terra Non Firma Earth*.[1] He described how scientists had measured an average upward (radial) motion of 18 mm per year using measurements from Very Long Baseline Interferometry (VLBI) stations, but because they believed in a Constant Diameter Earth they were confident the Earth could not be expanding. He explained that:

> Robaudo and Harrison considered this increase in Earth radius to be extremely large. They obviously did not consider Earth expansion when making this judgement, but instead compared this to values that were expected from areas undergoing crustal rebound during glacial melting, estimated at less than 10 millimetres per year.
>
> It is significant to note that Robaudo and Harrison **"expected that most VLBI stations will have up-down [radial] motions of only a few mm/year"**, and they then recommended that the vertical motion be **"restricted to zero, because** [they considered that] **this is closer to the true situation than an average motion of 18 mm/year"** (Robaudo and Harrison, 1993, page 54.[2])
>
> Robaudo and Harrison were, in fact, faced with a daunting problem. When they calculated the global geodetic network from 15 years' worth of observational data, they found, but failed to acknowledge, that the Earth was expanding by 18 millimetres per year.[3]

Maxlow then goes on to explain how this elimination of any increase in radius has manifested itself in large periodic adjustments in charts of the published data. One chart located near Canberra in Australia shows an arbitrary adjustment of 71 mm during 1993 to 1994. This would have resulted in a severe earthquake if it were real. Similar severe adjustments are also noted in a selection of other charts.

John K. Davidson, the geologist who was taught by Professor Carey, now runs a successful consultancy that uses innovative new geological techniques to help companies find and exploit oil

---

[1] *Maxlow, 2005.*

[2] *Robaudo and Harrison, 1993.*

[3] *Maxlow, 2005.*

Chapter 3 - Drifting Continents

fields on the ocean floor. His pragmatic view is that his new techniques 'can still be used in the absence of a cause' because they work, just like many new industrial processes that are generally in advance of scientific theory, but also theorises that his new techniques work because Earth expansion is reducing the surface curvature of the crust in a series of pulses. In support of this argument he has highlighted in various publications a number of inconsistencies with the analysis of Earth measurements. In one recent technical paper in the APPEA journal in 2008 he commented that:

> ... in 2002, Gerasimenko and Kasahara ... removed 23 of the 59 stations in the VLBI database because the measured vertical component of those stations exceeded 4 mm per annum and each was considered unreliable. This discrimination against 40% of the data caused a minor contraction of 0.3 ±0.1 mm per annum and preserved the status quo with plate tectonics under constant radius.[1,2]

The obvious point being made by Davidson is that the data only supports the Constant Diameter Earth if the 40% of the data showing radial expansion is ignored. If all the available data is used it indicates Earth expansion.

Further evidence for expansion is provided by Ilton Perin, who suggests that if a great circle measured the Earth's diameter without crossing any subduction zones then this *must* show if the Earth is a constant or expanding diameter. Using data provided by the U.S. Geological Survey, a suitable great circle indicated that the Earth's circumference is expanding today at an annual rate of 77.8 mm per year.[3] The Earth's radius needs to increase by 24 mm per year to achieve this expansion.

But despite all this evidence, discussion of the Expanding Earth is still almost totally suppressed in today's leading scientific journals and the only allowable concept is a Constant Diameter Earth.

Scientific papers continue to be published and discussed in more open-minded journals such as *New Concepts in Global Tectonics (NCGT)*, where creative ideas not fitting readily within the scope of Plate Tectonics are examined. Geologists have also continued to publish books discussing the evidence for the

---

[1] *Davidson, 2008.*

[2] *Gerasimenko and Kasahara, 2002.*

[3] *Perin, 2003.*

Dinosaurs and the Expanding Earth

Expanding Earth theory.[1] Craftsmen like Ben Berends have constructed physical Expanding Earth models that stretch the ocean floor to demonstrate how the original continental shell would naturally split and move the continents apart by radial expansion until the continents were positioned exactly where they are today.[2]

The similarity of ancient fossils found on continents as remote as Africa and South America was one of the main reasons Wegener had become convinced these continents were once joined together. The geologists of his day rejected his drifting continents theory as 'an absurd daydream'. Thirty years after his death, during the 1960s, geologists began to realise that he was basically correct and today it is accepted that the similarity of ancient fossils is because Africa and South America were once joined.

Various authors have highlighted evidence that similar ancient fossils, now thousands of miles apart on opposite sides of the Pacific, also indicate the Pacific Ocean did not exist on a smaller diameter Earth. David Noel was director of the Tree Crops Centre when he wrote *Nuteeriat: Nut Trees, the Expanding Earth, Rottnest Island, and All That ...* in which he looked at how the distribution of nut trees and other forms of life support the concept that the Earth has expanded.[3]

In the *Journal of Biogeography* Dennis McCarthy also analysed life around the Pacific and compared it with geological evidence in order to check the likely size of an ancient Pacific Ocean.[4] The study looked at both fossil and living life from East Asia, Australia, New Zealand, South America and North America that appeared to be linked across the Pacific. He also came to the conclusion that the Pacific was once closed on a smaller diameter Earth.

All these various strands of evidence reveal a revolution in our knowledge of the Earth. The Constant Diameter Earth theory needs to be replaced by an Expanding Earth.

It may seem that all science is a continuous and linear progression so such a revolution in science would be highly unusual. Thomas Kuhn was an American science historian who proposed in his book *The Structure of Scientific Revolutions* that

---

[1] Chudinov, 1998: Luckert, 1998: Hoshino, 1999: Carey, 2000: Bridges, 2002: Maxlow, 2005: Shehu, 2005: Findlay, In progress.
[2] Berends, 1996.
[3] Noel, 1989.
[4] McCarthy, 2002.

Chapter 3 - Drifting Continents

all science is much more dynamic than commonly supposed and regularly undergoes periodic 'paradigm shifts' of knowledge that have fundamentally revolutionised many different science disciplines.[1] He argued that science does not progress with a steady accumulation of new knowledge. Periodic intellectual revolutions where a particular science is abruptly transformed is the most common form of scientific progress. Because science text books only describe the current accepted theories, rather than the history of science, almost all of these past scientific revolutions are effectively hidden and this fosters the common belief in the steady progress of science.

Kuhn maintained that most average scientists tend to be conservative individuals who accept the theories they have been taught. They attempt to explain facts with the currently accepted theories and tend to ignore or even dismiss research that challenges these accepted notions. As a result, it is frequently the open-minded individuals not deeply indoctrinated into accepted ideas who often support the most radical new theories. These new theories provide an essential tension in science which stimulates progress.

A new theory can develop, sometimes over decades, which is fundamentally different from the accepted notion. Both theories are able to explain the facts so different scientists, using the same facts, can support one or another of the theories. Eventually a new theory can suddenly replace the old theory once the evidence becomes overwhelming. After a scientific revolution one conceptual world view is abruptly replaced by another that provides a new more complete understanding of nature.

Sometimes these revolutions are so major there is considerable resistance to the new theory resulting in a period of scientific inertia that can easily last for decades. This is particularly true when the old theory has become such a strongly entrenched scientific dogma that the new theory is hard to accept by the majority of scientists. It is clear this is the current stage of belief in the Expanding Earth theory.

In the next chapter I will explore these Expanding Earth concepts further by examining this expansion in size *and mass* in more detail.

---

[1] *Kuhn, 1962.*

4 - The Expanding Earth

The idea of an Earth which is constant and unchanging has been restated so often throughout history that it has now become established as a firm fact. It needs no proof - which is lucky since there is none.

There have been many instances where other beliefs have come to be treated as indisputable facts over generations and these beliefs have become so firmly rooted in the minds of mankind that it is sometimes dangerous to propose an alternative. A classic example was when the Polish astronomer Nicolaus Copernicus first suggested that the Earth orbits the Sun.

Between the years 1510 and 1514 Copernicus circulated within a small group of friends his manuscript summarising his radical theory that the Earth revolved around the Sun. This was a completely different view from that held by most people who believed that the Earth was the centre of the universe and that the planets, Sun and stars revolved around it. The idea that the Earth was the centre of everything was taken to be a firmly established fact since the Sun was observed to rise in the morning and move across the sky before it set at night. In the same way as the Sun, the moon and stars could also be seen to revolve around the Earth. For many, no further evidence was necessary.

Copernicus was wise enough to realise the anger that would develop when he aired his theory so he postponed publishing his book *On the Revolutions of the Celestial Spheres* until 1540 when he was near death.[1] A finished copy is believed to have been brought to him on the last day of his life.

One man who was later to support this theory - the famous Italian scientist Galileo Galilei - was not even born when Copernicus's book was published. Many years after

---

[1] *Copernicus, 1540.*

Chapter 4 - The Expanding Earth

Copernicus's death Galileo began to study the stars and planets through the newly discovered telescope, and what he saw amazed him. He saw Jupiter's moons and Saturn's rings, and these observations convinced him that the Earth and planets must orbit the Sun just as Copernicus had suggested.

Others refused to be swayed by any evidence. Galileo complained in a letter to Johannes Kepler in 1610 that some philosophers who opposed the reality of the observations had even refused to look through his telescope.

By the year 1613 Galileo's reputation was so well established that he published three letters in which he maintained that the Copernicus theory was correct. Although he was confident of his position and rank, he had misjudged the depth of belief in the Earth as the centre of the universe. The Aristotelian professors united against him and finally he was secretly denounced to the Inquisition for blasphemous utterances. Much alarmed by these events, he wrote letters to the Grand Duke reminding the Church of its standing practice of interpreting Scripture as conveying a deeper meaning than that on the surface whenever it clashed with scientific truth. This was not successful and resulted in the Copernican system being declared false in 1616.

While Galileo was arguing his beliefs with the Church, others were having more success. In 1609, Kepler had demonstrated that Mars revolves in an elliptical orbit around the Sun. Kepler was the first to extend Copernicus's reasoning to the other planets to show that all the planets were bodies like the Earth rotating around the Sun. By 1619 Kepler had published all three of his principles of planetary motion.

Meanwhile, Galileo was still restricted from adding his support to this new theory, but by 1624 a new Pope was in position and Galileo obtained permission to write about the two systems of the world - one with the Earth at the centre of the solar system and the other with the Sun at the centre. *Dialogues concerning Two World Systems* was published in 1632,[1] but by 1633, despite having obtained a licence from the church for the publication of his book, Galileo was prosecuted by the Inquisition on suspicion of heresy.

Heresy was a very serious crime. In 1600 the Italian authorities had burned the philosopher Giordano Bruno at the stake after the Inquisition had found him guilty of heresy for his cosmological theories. There was a very real possibility that Galileo would suffer the same fate.

---
[1] *Galilei, 1632.*

Galileo was found guilty of having held and taught the Copernican system but luckily he was only sentenced to imprisonment after he recanted the Copernican system. Due to his old age, the sentence was commuted by the Pope into house arrest although it remained in effect until his death eight years later.

Although Galileo spent the remainder of his life under house arrest he continued his work on mechanics and completed his book *Discourses and Mathematical Demonstrations Relating to Two New Sciences* which was the first to highlight the scale effects on life, as mentioned in the first chapter.[1] Since all his works were banned from publication, including any he might write in the future, the manuscript was smuggled to Holland to be published.

Fortunately for our own understanding of the stars and planets, the Copernican theory has developed into the basis of our modern theories about the universe. Today, when we can see pictures of the Earth floating in space, it may be easy to look on our ancestors as foolish. But if they were foolish to believe that the Earth was the centre of the universe, could we be even more foolish to commit a similar error by denying that the Earth could have expanded without first examining the evidence?

Confirming the Expanding Earth Theory

As I have already explained, soon after I had realised that the most obvious cause of an increase in gravity at the surface of the Earth was an Increasing Mass Expanding Earth, I found that the theory of an Expanding Earth had been proposed before by Earth scientists. The explanations for why this expansion occurred have been many and varied, but I believe the large size of prehistoric life restricts the possible explanations. When the dinosaurs began to dominate the land, the Earth was much smaller in both size *and mass*. The present-day continents were joined to form one large continent, but as the Earth's mass increased the continents began to split as new ocean floor filled the gap. Due to the Earth's expanding mass, its gravity increased and this gradually reduced the maximum size life could achieve.

I soon realised that if it were possible to determine an accurate size for the ancient Earth then this could be used to calculate accurately the gravity at the Earth's surface from 200 million

[1] *Galilei, 1638.*

Chapter 4 - The Expanding Earth

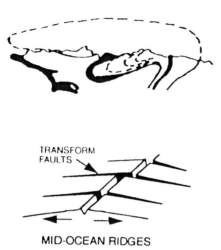

Fig. 4.1 Extension and Compression of the Earth's Crust

The Earth shows evidence of both compression and extension of the Earth's crust. Many mountains seem to be formed by the folding of the Earth's crust. In the ocean floor 'transform faults', which seem to be giant tears in the Earth's crust, are found running at right angles to ocean ridges indicating that the Earth is being torn apart in these places. Both compression and extension can be explained by an Expanding Earth.

years ago up to the present day. This variation in gravity could then be compared with the values calculated from the scale of prehistoric life. If the values of gravity obtained from the Reduced Gravity Earth theory and the Increasing Mass Expanding Earth theory agreed this would indicate that both theories were correct.

Using the idea that the continents of the Earth were once joined, I drew the continents onto a makeshift globe, then transferred the outlines of the continents onto a globe of about one half, and onto another one third the diameter. Using the geological evidence that Earth scientists had obtained for the age of the Earth at these times I was able to calculate roughly that the gravity at the Earth's surface agreed with the values I needed to explain the gigantic size of the dinosaurs.

It was then that I decided to obtain more accurate values. The magnetic recordings on the ocean floor have been mapped to give a detailed account of the age of the Earth's ocean floor. By removing the ocean floor that is known to be younger than a particular age, it is possible to reconstruct the ancient Earth by rejoining the remaining ocean floors. I had used this method to determine roughly the size of the ancient globe but transferring the continents onto a globe of smaller diameter tended to be inaccurate. The continents and their surrounding ocean floors were easily distorted as they were moved to the smaller globes so it was obvious I needed a more accurate method.

One much more accurate method is to first calculate all the positions of the outlines of the continents and ocean floors using spherical trigonometry on the Earth of today's diameter. Then, using an ancient Earth diameter which was smaller, the outlines of the continents and ocean floors can be recalculated on this smaller world. By moving and rotating these plates and then re-calculating their outlines it should be possible to adjust the diameter of the Earth until all the outlines mesh together.

There is one problem with this method of calculation which cannot be overlooked - if a section of a globe is cut and placed on a larger globe, the section will not fit without distorting it. There are two possible methods of distorting this section - either the outer edges of the section can be torn to fit the larger diameter, or the inner part of the section may be crumpled. I needed to decide which way the surface of the Earth had been distorted to perform the calculation, but which method should I choose?

In terms of deforming the Earth, both of these methods should leave clear signs in the Earth's crust. If the central section is crumpled and crushed it would tend to produce mountains by folding of the Earth's crust as shown in the top section of Figure 4.1. There are indications that most mountains are caused by just such folding actions. For the continental crust there is evidence to suggest that it has been crumpled by some immense forces. The amount of crumpling varies with the strength of a particular region. Although the expansion of the Earth may result in mountain building in areas where the continental crust is weak, the major portion of the continental crust remains solid despite the distortions imposed on it.

Fortunately the continents only cover a small portion of the Earth's surface. Most of the Earth's surface is covered by ocean floor, and this gives every indication of having been torn apart at the outer edges. On examining the newly-formed ocean floor at the ocean ridges it is possible to observe transform faults running at right angles to the ocean ridges as shown in the bottom section of Figure 4.1. These faults are kilometres wide in places. An indication of how these transform faults would occur on an Expanding Earth can be gained by imagining a section of any small sphere being placed on a larger sphere. The smaller section will not fit unless the outer edges of the small section is torn apart. This same effect would tend to tear the outer edges of the ocean floor to produce the transform faults.

Chapter 4 - The Expanding Earth

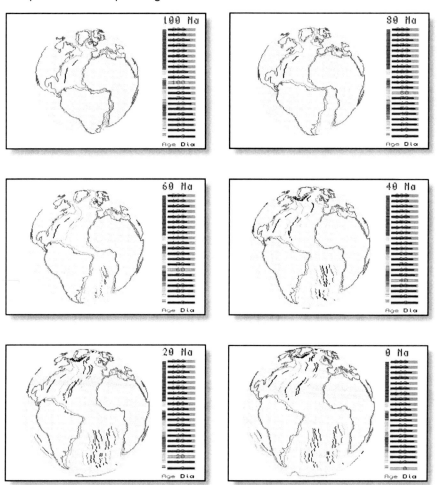

Fig. 4.2 The Author's Expanding Earth Reconstructions

Some of the Expanding Earth reconstructions produced by the author's computer program. Expansion is shown over the mid-Atlantic ridge in the examples above. The age of the Earth is indicated in the top right corner of the reconstructions in millions of years - so 20 Ma is 20 million years ago, 40 Ma is 40 million years ago, and so on. The lines on the ocean floor define the known age of the ocean floor, so 100 million years ago most of the ocean floor did not exist. As new ocean floor was created the continents slowly separated until they reached their present-day positions.

Dinosaurs and the Expanding Earth

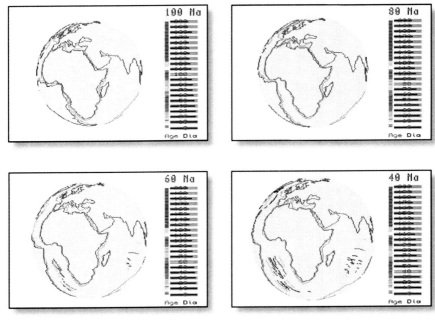

... more expanding Earth reconstructions this time over Africa. All these reconstructions only show the major movements of the ocean floor and continents with no major distortions or additions. In practice there have been many minor changes that are not shown. The Mediterranean has formed only recently within the last few million years. The African rift is separating the horn of Africa from the rest of the continent. North and South America have only recently been joined by the newly-formed Panama strip of land. See main text for details.

Chapter 4 - The Expanding Earth

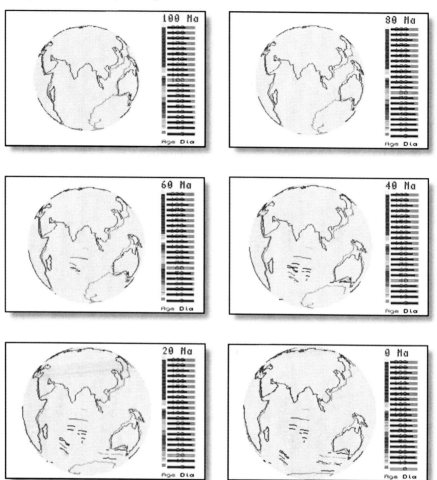

... more expanding Earth reconstructions this time over India.

Dinosaurs and the Expanding Earth

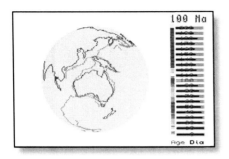

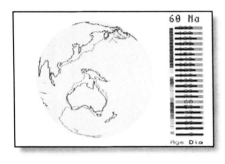
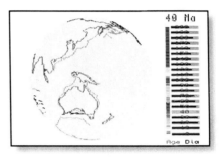
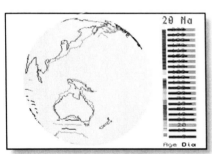
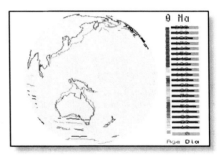

*... more expanding Earth reconstructions
this time over Australia.*

Chapter 4 - The Expanding Earth

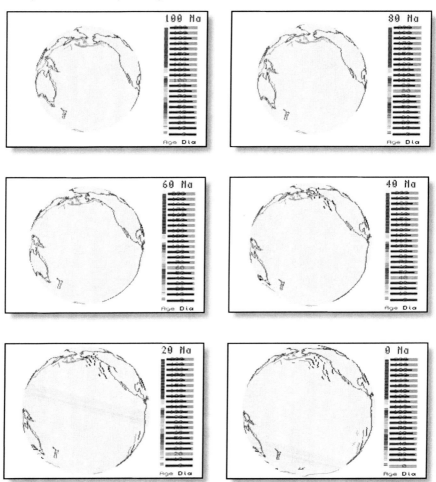

... more expanding Earth reconstructions this time over the Pacific.

Dinosaurs and the Expanding Earth

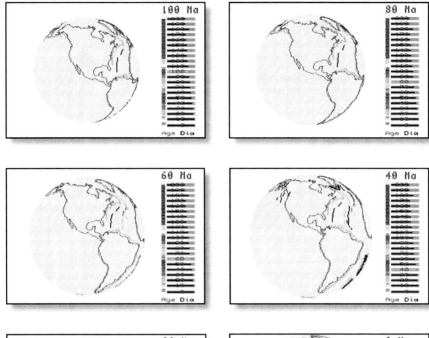

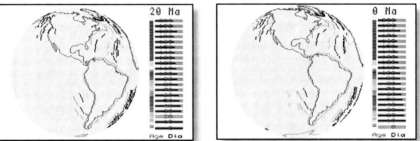

... more expanding Earth reconstructions this time over North America.

Chapter 4 - The Expanding Earth

The surface of the Earth therefore shows both forms of distortion. The continents, which are usually at the centre of an ancient section of crust, tend to show crumpling in the weaker part of the crust, while the outer portion of this section, which tends to be ocean floor, is torn and stretched at its outer edges. Since the continents are small compared to the ocean floors, I decided to calculate the ancient plates' size and shape by assuming that most of this distortion was caused by being torn at the ocean ridge.

This task was difficult because of the large number of calculations involved. A continent the size of Africa would need at least a few hundred complex calculations to outline its shape, and after repositioning it on a smaller globe, a further few hundred to outline it on a smaller globe. All the ocean floor ages on each continental plate also need to have similar mathematical treatments. Multiply these calculations by the number of plates on the globe, the movement of the plates, and the variation in size of the Earth with time, and you have an immense number of calculations.

With such a large number of calculations, the quickest way to visualise the ancient Earth was to program a computer. After a few years' work the program allowed me to reposition the various continental plates on a smaller diameter Earth. The maps can be redrawn in Mercator projection within a few minutes. The picture of a globe takes slightly longer. This quick redraw time allowed me to position all the major plates while varying the diameter of the Earth until the plates fitted neatly together with a high degree of accuracy.

One of the most useful plates for determining the size of the ancient Earth was the European-Asian-North American plate since this plate is effectively joined together to form one continuous continental land mass. The Continental plate stretches from Russia, across the Bering Sea, and into Alaska. Because this plate stretches completely around the globe, there can be only one diameter for any particular age. Having determined the correct diameter for the ancient Earth at 10 million year intervals up to 200 million years ago, the rest of the plates were fitted in position. A small number of these computer-generated Expanding Earth reconstructions based on geological data are shown in Figure 4.2.

Various other people have also produced reconstructions which look very similar to these models. By the third edition of this book, James Maxlow's reconstructions were widely available on the Internet and he had discussed the scientific basis for the reconstructions in detail in his book, *Terra Non Firma Earth*.[1] These reconstructions use the latest data for the age of the ocean floor. Neal Adams's reconstructions are also widely available on the Internet and elsewhere, as are various other reconstructions which have also been independently published over the years by people such as Ott Hilgenberg, Hugh G. Owen, Klaus Vogel, Karl W. Luckert, Ken Perry and others, which once again look similar. The fact that the reconstructions were produced without any assistance or reference to other reconstructions, and yet show very similar results, must give us all confidence that the results are substantially correct.

It is difficult to explain these reconstructions without using the Expanding Earth theory. One explanation, proposed by the American scientific geology writer Andrew Alden, is that we are all making the same mistake and that the Expanding Earth is an optical illusion.[2] It only looks like the Earth has shrunk to enable all the continents to fit together.

In practice there is a problem with this explanation. For the optical illusion to work the continents must have split apart at exactly the right positions. If the continents split in the wrong positions then there would either be large gaps or overlapping areas in the Expanding Earth reconstructions. But there aren't any. The reconstructions of a smaller diameter Earth are a perfect fit.

I calculate the chance of just one continental edge splitting in the correct place, so as to give such an optical illusion, is less than one in a million. Since all the continents need to split in exactly the correct positions the probability of this happening for all continents by chance is even less than this.

In addition, the Expanding Earth reconstructions can be progressively continued into the past. If we remove all the ocean floor that didn't exist 10 million years ago, the ocean floor and continents join together on a slightly smaller diameter Earth. The process can be repeated by removing all the ocean floor at all time intervals up to 200 million years old to slowly reduce the size of the Earth as it becomes more ancient. Eventually only the ancient continents are left and these can be joined together to form a continuous continental shell on an ancient smaller diameter Earth.

---

[1] *Maxlow, 2005.*

[2] *Alden, 2010.*

Chapter 4 - The Expanding Earth

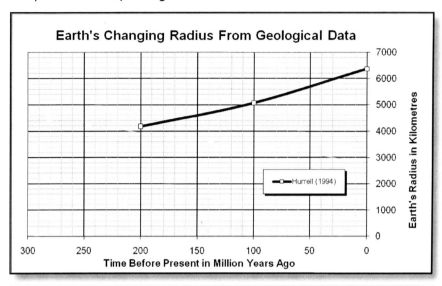

Fig. 4.3 Earth's Changing Radius from Geological Data
Reconstructing all the Earth's ancient continental and ocean crust determines the Earth's changing radius over periods of hundreds of millions of years. The author's estimate of the changing radius is shown based on the author's Expanding Earth computer models previously described.

Consider how unlikely it is that the entire ancient ocean floor fits together so precisely at all time intervals. If the missing ancient ocean floor had been generated by any other process than an Expanding Earth it would be impossible for the continents to be the exact shapes required to reconstruct a smaller Earth. The Expanding Earth must be a real geological process!

Since the various ages of the ocean and the land mass can be dated it is possible to calculate the diameter of the ancient Earth. The main assumption required is that the Earth has expanded and new seabed floor has been created to fill the gaps. The results of calculating these values are shown in graphical form in Figure 4.3. This clearly shows an exponentially increasing expansion in the size of the Earth since the dinosaurs roamed it.

Ancient Surface Gravity

The variation in the Earth's surface gravity, which explains the size reduction of life, can be calculated from the Earth's changing diameter determined from geological data, assuming that mass has increased to produce the expansion. This is the concept of the Increasing Mass Expanding Earth.

The laws of gravity go right back to Isaac Newton and the famous apple that first started him thinking about gravity. Newton's renowned book the *Principia,* outlining the laws of gravity, was published during the final year of his life in 1727.[1] The concepts in it are still used today since it defines the basic laws of motion and gravity that describe why planets stay in orbit around the Sun or why we don't fly off into space. They are used today in thousands of different ways and without them we would not have been able to even dream of doing some of the more amazing activities like landing men on the Moon.

Newton proposed that gravitational attraction is a property of all matter so that, for example, the Earth is attracted to the Sun and we are attracted to the Earth. The magnitude of this attraction force is proportional to the square of the distance between objects and the mass of each object. This can be represented in a formula as:

$$F = G \times M_1 \times M_2 / R^2$$

where M_1 and M_2 are the masses of the two mutually attracting bodies, R is the distance separating them and G is the Universal Constant of Gravity. The calculated force F is effectively the weight of the small body M_1.

If we are on the Earth then the mass of the Earth M_2 and the small body M_1 attract each other to produce the force of gravity at the surface of the Earth. If the mass M_1 at the Earth's surface was 1 kg the force of gravity acting on the mass would be 9.81 Newton.

$$F = G \times M_1 \times M_2 / R^2$$
$$F = 9.81 \text{ Newton}$$

The magnitude of this gravitational force depends on the mass of the planet acting on the mass of the 1 kg body.

Once we have calculated the diameter of the ancient Earth and assuming that the mass of the Earth has varied in proportion to its diameter, with no change in the Earth's density, we can use this formula to calculate gravity at the Earth's surface. If the diameter of the ancient Earth was half its

---
[1] *Newton, 1727.*

Chapter 4 - The Expanding Earth

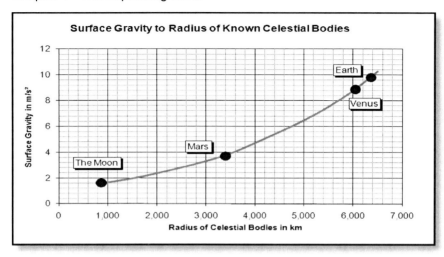

Fig. 4.4 Gravity Variation of Known Celestial Bodies
Existing celestial bodies can be used to produce a graph showing surface gravity in relation to radius with density changes included.

present diameter then its mass would be an eighth its present mass. This is because radius is linear but mass is cubic - it's the scale effect again but this time applied to planets. So the force acting on the 1 kg mass on a planet half the radius of the Earth would be:

$$F = G \times M_1 \times (1/8 \times M_2) / (1/2 \times R)^2$$
$$F = 4.905 \text{ Newton}$$

To put it another way - since G and M_1 don't change - gravity on a planet half the Earth's diameter, with an unchanging Earth density, would only be half the Earth's present gravity. Any object on the smaller diameter Earth would only weigh half its normal Earth weight.

The assertion that a smaller diameter, less massive ancient Earth would have less gravity is confirmed by our knowledge of smaller, less massive bodies like the Moon and Mars and is a result that would be expected. Smaller, less massive moons and planets have reduced gravity.

This is a simplistic method of calculating the force of gravity since it assumes that the density of the ancient Earth is exactly the same as the present Earth. It is much more probable that as the ancient Earth grew larger in size it would become more

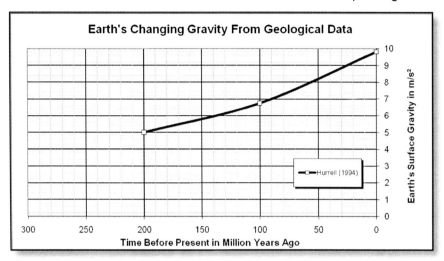

Fig. 4.5 Earth's Changing Gravity from Geological Data

Comparing the radius of the ancient Earth with the minor celestial bodies' radii gives an estimate of the Earth's changing gravity over time. Density changes are already included in the results using this method. Compare this increasing gravity to the changing gravity predicted from the scale reduction of life as shown in Figure 2.22.

dense as its core became more compact due to the increasing surface gravity.

This density increase would be difficult to quantify if it were not for the fact that there are other moons and planets in our solar system that are now the same size as the ancient Earth was in the distant past. By plotting the known variation of gravity against the radius of these celestial bodies we have a graph, as shown in Figure 4.4, from which the gravity of the ancient Earth may be directly read, since the radius of the ancient Earth can be estimated from geological data.

The final graph of changing gravity on the ancient Earth, taking account of density variations in the Earth's core and mantle, is shown in Figure 4.5.

You will note that the variation in gravity predicted from this Increasing Mass Expanding Earth is comparable with the variation that was also predicted from the scale reduction of life, as previously indicated in Figure 2.22, for the Reduced Gravity Earth. There are therefore two completely independent lines of reasoning which indicate that the Earth has expanded in size and mass to increase its gravity.

Chapter 4 - The Expanding Earth

Increasing Radius, Mass and Surface Gravity

Annual values of radius, mass and surface gravity increase can be calculated from the geological evidence for expansion. Although it is useful to define these annual values for comparison with other reconstructions it should be noted that, although the values are given in annual amounts, these results are average values based on a geological timeframe of millions of years. It would be wrong to interpret this as a true annual value for a period of time that is effectively an instant of geological time. The expansion of the Earth could change over geological time from periods of rapid expansion to periods of little or no expansion. I will return to this variation in later chapters when I discuss the reason for Earth expansion.

The annual values obtained from my computer generated geological reconstructions are:

Radius increase	= 14 mm per year
Surface gravity increase	= 3.6 microGal per year
	= 3.6 x 10^{-8} m/s^2 per year
Mass increase	= 1.7 x 10^{13} tonne per year

Various other published data, using a wide range of sources from Expanding Earth reconstructions to modern GPS data, can be interpreted as indicating radius increases of: 6 mm per year (Man and Kanagy), 12 mm per year (Carey 1958), 12 mm per year (Owen), 14 mm per year (Hilgenberg), 14 mm per year (Scalera), 18 mm per year (Robaudo and Harrison), 19 mm per year (Luckert), 20.8 ±8mm per year (Parkinson), 22 mm per year (Vogel), 22 mm per year (Maxlow), 24 mm per year (Perin), 32 mm per year (Carey 1988).[1] This range seems large but it should be compared to the general view that expansion is zero, so any expansion at all is considered unfeasible by geologists who believe in a Constant Diameter Earth. Also some methods calculate the expansion over short timescales while others use much longer ones, so the variation in the results may simply confirm that the rate of expansion is increasing. No doubt more consistent expansion results will be obtained in future when more resources and finances become available.

---

[1] *Davidson ,1994 (Man & Kanagy): Carey, 1958: Owen, 1983: Hilgenberg, 1962: Scalera, 2003: Robaudo and Harrison, 1993: Luckert, 1999: Parkinson, 1988: Vogel, 1990: Maxlow, 2005: Perin, 2003: Carey, 1988.*

The Earth of 300 Million Years Ago

It is now possible to reconstruct the early world of 300 million years ago which was very different from the present. Most of the ocean floor did not exist. As described in previous chapters, instead of being split into several continents separated by vast stretches of ocean there was only one super-continent. Much of this super-continent was covered in shallow seas. This super-continent, or Pangaea, was composed of the major land masses of the present world. The world itself was much smaller, being about half its present diameter.

Due to the Earth's smaller size and mass the gravity at its surface was much less than at present. Because of this low gravity life evolved to reach gigantic sizes (compared with present-day life), with giant dragonflies, millipedes, horsetails, club mosses and amphibians larger than present-day crocodiles. The larger scale of life like the dinosaurs had not yet developed and so these giant insects and amphibians were the largest life of 300 million years ago.

There have been many gigantic fossils found from this time which show that life did reach gigantic sizes, but there should logically be other effects caused by the low gravity. One effect would be to allow mountains to become much higher than they are now. Present-day mountains are limited in size because if they become too large the stresses at their base become too high to allow them to grow any bigger. A lower gravity would therefore allow these mountain-building forces to push the mountains higher than those of today.

Uniform Temperatures

The climate in Carboniferous times was uniform over most of the Earth and about 300 million years ago even the poles were warm. This is demonstrated by the fact that the land masses known to have been at these polar regions now contain the coal of an ancient forest. The polar regions lacked ice caps and supported a variety of life which has left fossils in the now cold regions of Greenland and Antarctica. The warm conditions of the tropics extended to form a wide warm band into latitudes as far as 60° from the equator. The warm conditions allowed great tropical reefs to extend this far and leave tropical fossil reef corals in marine deposits.

Chapter 4 - The Expanding Earth

Because of the equal range of temperatures over the globe, the animals of these times tended to roam across the whole world since there were no physical or climatic barriers. It is therefore difficult to distinguish any difference in the fossils of this time. The same fossil plants have been found in north-eastern Europe and the eastern Mediterranean, and Nova Scotia and central Kansas. This is true of both land and marine life.

Another indication of the uniform temperature over the Earth in Carboniferous times is shown by the absence of annual growth rings in coal-swamp trees. It seems that there were no seasons to produce variations in temperature and rainfall. On the present-day Earth, the distinct seasons produce pronounced changes in the rate of growth of these rings.

Why the Earth's warm temperature extended from the equator to the poles has been as much of a mystery as the giant size of life in those times. But with an Earth of smaller diameter, the climate at the poles would be naturally forced to become warmer by the heat flux of the atmosphere over the Earth.

The atmosphere traps the heat by the greenhouse effect. The energy from the Sun enters the atmosphere in the visible part of the spectrum and warms the land or sea. These then re-emit the heat as infra-red radiation which is now absorbed by the water vapour and the carbon dioxide in the atmosphere. The heat is effectively trapped by the lower layer of the atmosphere known as the troposphere.

The atmospheric circulation of the Earth's weather system is driven by the heat from the Sun which warms the air and water vapour in the equator regions so the hot air rises converting the heat energy into the kinetic energy of atmospheric circulation. The speed of air movement is limited by the friction of the wind against the Earth's surface and this friction limits the amount of heat which the atmosphere can transfer from the equator to the poles with an almost linear fall in temperature from the equator to the poles. Given that temperature fall is constant with distance, the effect of a smaller Earth would be to reduce the temperature difference from the equator to the poles. This fact would effectively increase the amount of heat which could be carried to the poles by atmospheric circulation so raising the temperature of the poles. Also, the increase in temperature can be calculated from the temperature drop over distance and the size of the Earth over millions of years.

A Model of Uniform Temperatures

Although it would be extremely difficult to predict the change in weather patterns on the Expanding Earth in great detail, it is possible to construct a simple model of the Earth that shows the temperature conditions present in the past. This simple model Earth can be considered to be a planet which orbits our Sun and receives the same amount of radiation from the poles to the equator as the real Earth. This model Earth has no day and night since it does not revolve. It has no mountains and no seas or continents - its surface is totally smooth. It does have an atmosphere, and this is warmed by the heat from the Sun being

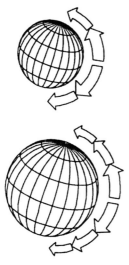

Fig. 4.6 Changing Climatic Belts
The atmosphere transfers heat from the equator to the poles. A smaller diameter Earth would tend to have warmer poles with less temperature variation with the seasons. The climatic belts would have moved as the Earth expanded. All these conclusions are supported by geological evidence.

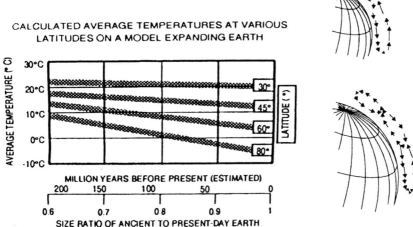

Chapter 4 - The Expanding Earth

reflected from the planet's surface. This atmosphere tends to carry the heat from the equator to the poles so that the model world shows the range of temperatures of the real Earth. The equator and poles are at the average temperature of the real Earth, but the temperatures never vary because this model Earth has no day or night, no winter or summer, and no mountains or seas to disturb the flow of heat carried by the atmosphere.

This model Earth can also vary its diameter. This allows the effects of an Expanding Earth on the heat flux in the atmosphere to be quantified. Using this simple model it is possible to give an average value of the temperatures at various places on the Earth's surface for an Earth of varying diameters. The variation in average temperature over time at various latitudes can therefore be plotted in graph form as shown in Figure 4.6.

The Earth of 300 million years ago was therefore very different from the present due to its smaller diameter and mass, with all the present continents joined. Due to the smaller mass of the Earth the gravity was weaker and this produced several effects. The insects, amphibians and plants of the time were able to achieve gigantic sizes. The mountains should also have achieved a larger size. Life roamed all the continents not only because they were joined but also because the temperature was more even due to the Earth's smaller size.

The Splitting of the Continents

As the Earth expanded, the original continental shell which covered the whole Earth began to split first in the Pacific Ocean region, while the continents remained connected across what was to become the Atlantic Ocean until well into the Jurassic period, about 150 million years ago. Then a narrow ocean began to form between Africa and North America as new ocean floor filled the gap. This ocean became wider throughout the period but it did not finally split the two continents until the Cretaceous times, about 100 million years ago.

The sea level in the early Mesozoic showed marked changes. It remained relatively low during most of the Triassic and early Jurassic but then increased rapidly during the late Jurassic until a quarter of the present continental area had been flooded. Despite this rise in sea levels, the land dinosaurs must have continued to migrate between the continents as the fossil remains still show strong similarities between the Western United States, Argentina and East Africa.

In the journal *Science and National Geographical Research* palaeontologists described the conditions at the poles. By 130 to 105 million years ago in the early Cretaceous period a range of small or medium-sized dinosaurs lived near the South Pole. The fossils from Australia, which was then attached to Antarctica, show that the average temperatures were 5°C (41°F) and could fall to - 6°C (21°F). The summer months were wet and cool with plenty of plants. Even during the winter months, while the dinosaurs saw no sun, there was plenty of vegetation. There was a variety of evergreen herbaceous plants and trees amongst many types of dinosaurs. Australian palaeontologists identified species of plant-eating and meat-eating dinosaurs, pterosaurs, freshwater reptiles and turtles.

The continents we know today began to form in the late Mesozoic, as Earth expansion continued. North and South America broke away from Africa and Europe to form the Atlantic Ocean. In the late Jurassic and early Cretaceous the continents flanking southern Africa broke away, forming the Indian and South Atlantic Oceans. As South America rotated away from Africa, the South Atlantic pushed northward between them until it finally split them in the mid-Cretaceous period. At the same time, North America continued to separate from Europe, creating the North Atlantic.

The warm conditions and broad climatic belts still extended over much of the world because it was smaller than now. It seems unlikely that there were any permanent polar ice caps at this time, although there may have been spells of colder weather when ice formed.

The conditions appear to have become more humid during the Cretaceous. Again in the journal *Science* other palaeontologists reported that in the late Cretaceous period, between 80 to 65 million years ago, other dinosaurs lived in northern Alaska in cold and wet conditions near the North Pole at between 70°N and 85°N. These dinosaurs occupied the polar regions for at least tens of thousands of years. This was a great contrast to the arid conditions experienced during the Jurassic.

During the late Mesozoic angiosperms (flowering plants) evolved. This must have changed the appearance of the forests to give them a distinct modern look, but there was still no grass. In the seas newly-evolved algae were laying down a new sea floor of calcite, which now formed the chalk of much of northern

Chapter 4 - The Expanding Earth

Europe and Texas. At the very end of the Mesozoic there was a very rapid fall in sea level which turned many marine areas into land. The close of the Cretaceous about 65 million years ago was marked by many extinctions, both in the sea and on land.

During the Tertiary, starting about 65 million years ago, the continents continued to move towards their present positions and more ocean floor formed between them. In the south, Antarctica continued to move away from its adjoining continents. The last continent to break away from Antarctica was Australia, which stayed joined until the late Palaeocene about 55 million years ago. The two Americas were isolated from each other until the Panama land strip reconnected them about 3.5 million years ago. In North Africa, it appears that heating of the crust caused it to distort into a dome-shaped structure until it eventually caused several rifts to appear there. This process also produced the Red Sea, the Earth's youngest ocean. The Gulf of Aden formed along a line connecting the Red Sea to the Indian Ocean and a third arm created the East Africa rift system.

The generally warm conditions that the dinosaurs had experienced towards the poles during most of the Mesozoic Era started to disappear with the rise of the mammals in the Tertiary Era. As the Earth began to approach its modern diameter, the climatic belts became more bunched around the Equator. Glaciation at the poles began about 40 million years ago in the Eocene Era and by 30 million years ago in the early Oligocene sea ice had started to appear. Then, about 5 million years ago, at the end of the Miocene, the Great Ice Age began to extend the glaciation over a third of the world.

The Mediterranean was created as Europe moved away from Africa, but it became isolated from neighbouring oceans and virtually dried out by evaporation to leave vast layers of salt beds. This lasted for about one million years until the Straits of Gibraltar formed to allow the sea to refill from the Atlantic.

Today the Earth has many scars which show how it was forged by geological activity. Under two kilometres of ocean are hidden the mountains and valleys which emerge from the repeated eruption and cooling of molten lava. This is the spreading zone of the ocean crust where the sea floor separates. It marks the centre line where the continents move apart as the Earth expands.

Set high on the mid-Atlantic ocean ridge sits Iceland. The volcanic action of the ridge sleeps fitfully and sometimes breaks out in Iceland itself. Some faults also form as the plates of the

Earth slide past each other. The most famous of these is the San Andreas Fault, and movement along this fault is bringing Los Angeles and San Francisco together as the plates slip relative to each other.

The largest rift valleys lie under the ocean floor, but the action can be seen at some places on the Earth's crust. The Red Sea and the great African rift were formed as the Earth's crust was torn apart. The Rift system extends from the Red Sea, through the Mediterranean, and down into Africa, where it is forming the great rift valleys.

During the Triassic-Jurassic transition about 200 million years ago, a new coral reef community evolved in southern Europe. Today, the remains of these large reef structures now stand exposed in the Alps, the sea floor having been raised up by mountain building movements, and the whole continental block crumpled by the Expanding Earth.

There is no doubt in my mind that the Earth has increased both its mass and diameter in expanding. But where has the extra mass come from? This is the subject of the next chapter.

5 - Meteorites and Ice Ages

In the last few chapters I have explained why I believe the Earth has doubled in diameter and increased in mass by about eight times, during the last few hundred million years. In this chapter I shall examine three possible causes for this expansion of the Earth with extra matter:

- Firstly, the interior core of the Earth is creating new matter,
- Secondly, cosmic particles are becoming embedded in the core of the Earth,
- Finally, comets, asteroids, meteorites and cosmic dust have repeatedly bombarded the Earth.

All of these three possible causes of Earth expansion would produce an Increasing Mass Expanding Earth. We can rule out some previous suggestions for the cause of Earth expansion since any concept that does not predict an ancient Reduced Gravity Earth cannot be the correct solution. Due to the scale reduction of life we know that the mass of our planet must have been growing to increase the Earth's surface gravity, so any process that produces a Constant Mass rather than an Increasing Mass Expanding Earth can be discounted.

New Matter

The first idea, that new matter is being created inside the core of the Earth, is outside any known physical laws acting in our universe. It is generally treated with suspicion, since matter has not been observed to come into existence.

Of course, the fact that this has not been observed to occur on the surface of the Earth does not mean that it is not occurring

Chapter 5 - Meteorites and Ice Ages

deep within the core of the Earth. The Earth's core is subject to great pressures and densities, so these unusual conditions may produce unique physical reactions deep within the heart of the Earth. But it must remain pure speculation since there is no direct evidence for, or against, this idea. Having stated this I tend to ask myself where matter began. Did it really come into existence at the beginning of time and space, or is it being created now?

One well-known proponent of new matter creation inside the Earth was Sam Warren Carey, the Tasmanian Professor of Geology, who proposed that mass creation is related to the expansion of the universe. He proposed that matter and energy are opposites of the same effect and are created together from nothing. They can also cancel each other out. Our universe is a 'null' universe continuously creating new matter; as matter passes beyond the knowable universe new matter must be created to keep the universe in balance. This seems to me to be similar to the continuous creation universe, suggested by the British astrophysicist Fred Hoyle, that was popular until the Big Bang concept won more general acceptance. A complete explanation of Carey's concept can be found in both his books, *Theories of the Earth and Universe* and *Earth Universe Cosmos*.[1]

Carey realised the dangers of trying to provide an explanation for the expansion of the Earth and reminded readers in his book *Theories of the Earth and Universe*:

> ... if the explanation I offer should turn out to be invalid, that explanation should be rejected, not the reality of expansion. Let us remember that Wegener presented his empiricism of continental displacement together with his physical explanation. When the latter was found inadequate, the whole idea of continental displacement was rejected for several decades, even though in the long run it was found to be essentially correct.[2]

Unfortunately, most people ignore this warning and habitually reject the evidence for Earth expansion simply because no one has yet offered a convincing explanation of *why* the expansion is happening. Geological evidence such as the global Expanding Earth reconstructions, and the unlikelihood of this occurring purely by chance, are simply ignored.

---

[1] *Carey, 1988, 2000.*

[2] *Carey, 1988.*

Another supporter of mass creation in the Earth's interior is Neal Adams who has independently followed the developments of the Expanding Earth theory since the 1960s. Adams theorises that the creation of matter within the cores of all celestial bodies is an ongoing process. He believes that the known nuclear process of Pair Production produces the spontaneous generation of new particles of matter within the core of the Earth. Since the Earth is not just expanding but also growing in mass, he calls his theory the Growing Earth.

Perhaps the best perspective on the present-day level of knowledge about the Expanding Earth was suggested in *New Scientist* by Hugh G. Owen in his article entitled *The Earth is expanding and we don't know why*.[1] His article urged us to remember that we are all still struggling to understand the implications of the evidence for Earth expansion.

Cosmic Particles

The second idea, that cosmic particles are becoming embedded deep within the core of the Earth, is based on the observation of the cosmic particles that bombard the Earth. The idea of matter being able to enter the Earth from space is similar in concept to that proposed by Hilgenberg in 1933. In his book *Vom wachsenden Erdball*, he suggested that an aether-like material from space enters the interior of the Earth to increase both the mass and volume of the Earth.[2] We no longer use aether as a scientific description, but the more modern concept of spacetime, where space and time are considered to be a single entity, could be considered to be similar to the old concept of aether. One interesting result of aether flowing into the Earth is that the flow would create surface gravity and this gravity would be equivalent to the inertia force of acceleration. Unfortunately I have been unable to obtain a copy of his book, and even more recent reprints of his work are unclear, so I cannot say if he also considered the effect this new mass would have on the surface gravity of the Earth. In any event, the physics of the day stated that this hypothesis of mass increase was physically impossible and the idea was dismissed.

Although we can now point to cosmic particles as a possible cause for the Earth's expansion in mass and volume, the

---

[1] *Owen, 1984.*
[2] *Hilgenberg, 1933.*

Chapter 5 - Meteorites and Ice Ages

amount of cosmic particles which are known to hit the Earth seem far too low to account for the new mass.

Could it be possible that the number of cosmic particles vary in time and space? This seems unlikely from the current theories of what causes these cosmic particles, since the whole of the interstellar space in our galaxy is thought to be filled with vast numbers of these particles. They appear to come from every region of the sky and so no one source can be pinpointed.

There have been suggestions from several scientists that there may be forms of elementary particles of a kind not yet detected. Until recently these particles could be almost anything, but studies with particle accelerators have set new limits. These experiments have shown that varieties of the neutrino elementary particle are three known types plus an intriguing half a neutrino. It is this half which is causing speculation. Some theorists speculate that a particle about 10 times the mass of a proton may exist, but many experimenters are developing detectors to search for particles with masses lower than 10 proton masses.

There are three major reasons I cannot accept any of these explanations for Earth Expansion. Firstly, they do not provide any explanation for subduction. Subduction zones are areas of high volcanism, earthquakes and mountain building and any theory about the Earth needs to include an explanation for these facts. Secondly, newly-created matter would need to match the current velocity of the Earth otherwise the Earth's obit around the Sun would be rapidly disrupted. Thirdly, close study of most Expanding Earth reconstructions shows that expansion is most rapid across the oceans and this seems to indicate that the cause of expansion is related to oceans in some way. As I will explain shortly, there is one theory that can explain all these facts within an Expanding Earth model.

Meteorite Bombardment

If the last two ideas do not seem able to account for the extra mass added to the Earth, the last idea provides a wealth of facts to convince me that the Earth's mass has been increasing by the continuing, but somewhat intermittent bombardment of cosmic asteroids, comets, meteorites and dust.

Over geological time the Earth has been continually bombarded so it has been involved in a constant process of accretion (or accumulation) of new cosmic material. Small

Dinosaurs and the Expanding Earth

amounts of new cosmic material regularly collide with the Earth as 'cosmic showers' but over geological time there is evidence that much larger 'cosmic storms' add vast amounts of new mass to the Earth. The actions of weathering and subduction then move this new mass into the interior of the Earth on timescales that are unimaginably long, with processes such as erosion, sediment transport and subduction that can be observed today.

Accretion of meteorites and asteroids has been proposed by a number of other people to account for Earth expansion. Carey mentions F. Dachille, S.V.M. Clube and W.M. Napier, L.S. Myers and others as supporters of this concept even though he did not believe this could be the primary cause of Earth expansion.[1] Carey rejected accretion mainly because he believed the rate of cosmic bombardment would reduce over time and this did not agree with the geological evidence for an exponential increase in expansion. But the latest evidence available indicates that there is continuous cosmic bombardment that would produce an exponential increase in the Earth's diameter.

Lawrence S. Myers contacted me in 2004 to discuss the concept of cosmic accretion and also sent me a copy of his monograph outlining his views.[2] He noted that:

> External accretion of extraterrestrial mass is irrefutable. Everyone knows about meteors and meteor showers that regularly enhance the night skies at certain times each year. Meteorites, the solid remnants of meteors that land on Earth, are also known to almost everyone, even though few may have actually seen one.
>
> Every meteoroid (they come in all sizes, from small particles, to pebbles, to small rocks, and megaton meteorites) striking Earth's atmosphere at night creates a visible luminescent 'shooting star' trail that indicates frictional ablation during its transit of the atmosphere. This is a visual signal of mass being added to Earth's surface. Whether or not a meteorite, or just its ablated dust particles, reaches the ground depends on its original size, its molecular composition, its angle of entry, and the depth and density of the atmosphere.[3]

---

[1] *Carey, 1988 - see page 326.*

[2] *Myers, 1983.*

[3] *Lawrence S Myers personal communication, 2004. (See also www.expanding-earth.org)*

Chapter 5 - Meteorites and Ice Ages

In a 1972 paper Myers coined the term Accretion Theory (with a new word accreation) to define the creation of the Earth from external accretion and has continued to defend his theory in various publications since then.[1] His latest paper outlining the concept, *Accreation - A New Theory of Planetary Creation (How and Why the Earth is Growing and Expanding)*, was published in 2010.[2] The Accreation Theory marries external accretion of cosmic material with internal expansion of the Earth's core as the interior heats up. Inclusion of internal core expansion is a major part of Myers's Accreation Theory since he believes this heating of the core is responsible for all present surface anomalies on the Earth including volcanoes and mountain chains. He turned 90 in 2010 but his next project is to finish a comprehensive book entitled *Accreation of the Earth and Solar System* which should be available soon.[3]

Even those scientists who do not believe there is sufficient cosmic bombardment to cause Earth expansion generally have little doubt that cosmic accretion caused the Earth to form at some point in the past. They believe that the Earth began as a collection of meteoroids and asteroids orbiting the Sun about 4,600 million years ago. Then, over the next few hundred million years the gravitational attraction of these bodies drew them together to form a small planet. As other bodies approached this ancient Earth small and large meteorites crashed into it, slowly building up our planet towards its present size. The theory is known as the Kant-Laplace Nebular hypothesis and is now well established as the leading theory to explain the Earth's formation.

Our ancient Earth must have been as marked with craters as the present-day Moon or Mars. Then, after eons of time, the atmosphere and the geological activities of the Earth processed and separated the ancient cosmic material. But despite these changes in the compounds of the meteorites, geologists are still able to observe that the broad chemical composition of the Earth is similar to the meteorites from which it is formed.

The only other major theory of how the Earth might have formed is that the Earth (and all the other planets and moons) were drawn out from our Sun by a passing wandering star. This theory has been discounted for several reasons, the major one being that geologists have shown that the elements contained

---
[1] *Myers, 1972.*

[2] *Myers, 2010.*

[3] *Lawrence S Myers personal communication, 2010.*

within the Earth have never been totally molten. From this fact it must be concluded that the Earth as a whole was never part of the molten interior of any star.

The Timing of Formation

If we are left with only one possible explanation of how the Earth formed, the *when* is a completely different matter!

The Earth is supposed to have formed within about the first fifth or less of its total age, before any life had begun. The theory goes something like this: at the very beginning of the Earth's formation it rapidly grew by meteoric bombardment up to its present-day size. Then the bombardment stopped. It stopped so completely that the Earth remained a constant diameter for about 4,000 million years as illustrated in the graph in Figure 5.1. Then at the end of the Earth's development the atmosphere rapidly evolved and life formed on this Constant Diameter Earth.

This timing of events provokes several difficult questions. Why should the Earth form within the very first hundred million years of its life and then stop? And why did the formation of an atmosphere take so long?

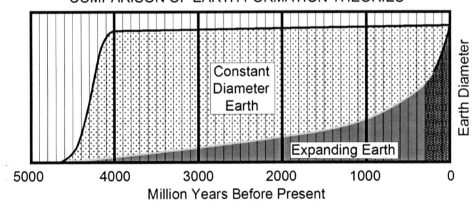

Fig. 5.1 Earth Formation Theories

Current Earth formation theories, like the Nebular Hypothesis, speculate that the Earth's size has remained constant for the last 4,000 million years. The Expanding Earth theory suggests that the Earth has formed gradually as shown in the shaded portion of the diagram. The more heavily shaded portion on the right uses the same geological data already shown in Figure 4.3.

Chapter 5 - Meteorites and Ice Ages

There seems to be no particular reason to believe this timing of when the Earth was formed. There is no startling fact that has been uncovered to prove that the Earth has been a constant size for thousands of millions of years. But despite this, only a few Earth scientists have ever suggested that the Earth may have expanded in the later part of its life.

Is the current timing of the Earth's formation a reasonable sequence of events? Or is it more reasonable to believe that the Earth began to form about 4,600 million years ago and has continued to increase in size since that time, as shown in Figure 5.1, in an ongoing uniform manner?

As we saw in Chapter 3, the idea of the Expanding Earth has had an interesting history, with some scientists supporting the idea to this day. In the previous chapters I explained why I believed this expansion was accompanied by an increase in gravity at the Earth's surface. The most obvious conclusion is that repeated bombardments of cosmic dust, comets and meteorites caused the Earth to increase its size, mass and the force of gravity over geological time - an Increasing Mass Expanding Earth.

This is still the same basic Kant-Laplace Nebular hypothesis of the Earth's formation but one where the creation of Earth is a continuous and ongoing process. It is an Ongoing Cosmic Accretion hypothesis for the Earth's formation.

Meteorites, Asteroids and Comets

There are many objects in outer space that are attracted to the Earth by gravity. These objects range in size from large comets and asteroids hundreds of kilometres across down to minute particles of dust micro-metres in diameter. All these have regularly entered the Earth's atmosphere from outer space to add additional mass to the Earth over an immense period of geological time.

Meteorites have been observed to enter the Earth's atmosphere in swarms that are not regular but appear to come in waves like 'cosmic showers'. But the estimated yearly average is only about 19,000 meteorites over 100g (4 oz) - hardly enough to increase the size of the world in a reasonable time.

By far the most cosmic material comes to the Earth as small dust particles. Scientists have estimated the accumulation of cosmic dust might amount to as much as 300,000 tonnes per year, but it's difficult to measure and no one is completely sure of the exact amount. Even this quantity is not high enough for our purpose.

But is it possible that we are at present in a quiet period? Could the Earth have entered a cloud of cosmic material that was much denser in the past?

When the first edition of *Dinosaurs and the Expanding Earth* was published in 1994 the general view was that not many cosmic bodies could hit the Earth. But various studies since then have provided some startling results that disturb this pleasant view. Increased interest in near-Earth objects has resulted in a few thousand near-Earth objects including asteroids, comets and meteoroids large enough to be tracked being identified.

Near-Earth objects orbit the Sun in a band so close to the Earth's orbit that they could hit the Earth. The United States Congress has mandated NASA to catalogue all near-Earth objects larger than 1 km (0.6 mile) wide in a project called *Spaceguard* and by 2008 they had identified 982 near-Earth objects.

If the search is widened to include near-Earth objects smaller than 1 km wide the amount increases dramatically. By 2010, 6,696 near-Earth asteroids and 84 near-Earth comets had been identified, and of these 1,086 have been classified as Potentially Hazardous Asteroids because they are very likely to hit the Earth at some point in the future.

Some asteroids have now been visited by spacecraft. The NEAR Shoemaker probe orbited the asteroid Eros to photograph its surface, as shown in Figure 5.2, and then eventually landed

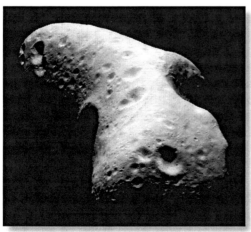

Fig. 5.2 Eros Asteroid

The NEAR Shoemaker spacecraft probe photographed the Eros Asteroid in 2000.
© NASA

Chapter 5 - Meteorites and Ice Ages

Fig. 5.3 Comet Impact

Many meteorites and comets must have been captured by the Earth's gravity over hundreds of millions of years.

on its surface. Eros is an asteroid 34.4 x 11.2 x 11.2 km (21 x 7 x 7 miles) in size. The mass of this one asteroid is much larger than all the cosmic material that is generally believed to have hit the Earth in recorded history - the asteroid Eros has a mass of 6.69 x 10^{12} Tonne - that's 6,690,000,000,000 Tonne.

Eros is the second largest near-Earth asteroid we know about. The asteroid Ganymede is the largest at 3.3 x 10^{13} Tonne, so it's roughly 5 times more massive than the asteroid Eros.

If such large asteroids hit the Earth head on they would create devastation on a global scale. But fortunately such a head-on collision seems remote.

Even the better option is not good. In practice an asteroid is more likely to follow the sequence of events of Comet Shoemaker-Levy that broke apart before it collided with Jupiter in 1994. The comet was probably captured by Jupiter's gravity in the early 1970s when its gravitational field pulled the comet apart as it orbited the planet, until it was a series of fragments that eventually hit Jupiter at a speed of approximately 60 km per second (37 miles per second). The size of the fragments ranged in size up to 2 km (1.2 miles) in diameter and produced at least 21 distinct impacts spread over 6 days.

If asteroids and comets are captured by the Earth's gravity as depicted in Figure 5.3, they would tend to break up as they approached the Earth and then spiral down in smaller-sized bodies. Many smaller bombardments could easily have occurred in prehistory and been incorporated in myth and ancient accounts of destruction.

Could it be possible that the Biblical account of the destruction of Sodom and Gomorrah was a real description of how the cities of the plain were destroyed by the remains of an asteroid or comet? The King James Version of Genesis 19 (24) recounts how as Lot and his family fled from the city of Sodom:

> Then the LORD rained upon Sodom and upon Gomorrah brimstone and fire from the LORD out of heaven;
>
> And he overthrew those cities, and all the plain, and all the inhabitants of the cities, and that which grew on the ground.[1]

Brimstone and fire would be a very good description of an asteroid hitting the Earth as it broke up into smaller fragments. Do we need to re-examine ancient accounts of destruction with a more open mind?

The meteorite bombardment of today is not enough to account for the formation of the Earth, but suppose that this meteorite bombardment is not constant but comes in waves. We know that there are 'meteorite showers' on a yearly timescale. Could there also be 'cosmic storms' on a geological timescale? Could it be possible that the amount of material from outer space has been much, *very much*, greater on the ancient Earth?

In fact, there is ample evidence that the Earth has been bombarded by vast amounts of cosmic material over geological time. The amount of material involved is so great that it has been related to the mass extinction of whole species of animals. A sudden increase in the amount of cosmic material has been blamed for one of the most well-known mass extinctions - the death of the dinosaurs.

The dinosaur mass extinction at the end of the Cretaceous, about 65 million years ago, affected many animals and plants, not just dinosaurs. The sudden disappearance of whole species occurred on land, air, and sea. These occurred when the Earth

---

[1] *Genesis 19 (24).*

Fig. 5.4 Impact Crater
Typical meteorite impact crater.
© *NASA*

Chapter 5 - Meteorites and Ice Ages

appears to have been bombarded with several large comets or asteroids.

There are several large craters known to have been created at the end of the Cretaceous and typical of these craters is the Manson crater in Iowa with a 40 km (25 mile) diameter. But could this crater be the result of only one of many impacts that occurred at this time? It would be easy to imagine that the smaller craters, like the one shown in Figure 5.4, would be obliterated by vegetation and erosion within a few thousand years.

Accompanying the extinction of life at the end of the dinosaurs' reign are fine intersecting lines within mineral grains of ancient rocks. These have been taken to be caused by intense shock, perhaps the intense shock of meteorites repeatedly hitting the Earth. Along with these indications, tiny spherules of rock and a concentration of what appears to be soot are found within the sedimentary layers that formed at the end of the Cretaceous.

All these signs are consistent with an impact of one or more massive asteroids or comets. After the collision, the trailing dust clouds and the dust thrown into the atmosphere would lead to darkened skies and a global reduction in temperature. It has been proposed that it was not the immediate effect of the impacts that caused the world-wide death of so many species. This has been blamed on the after-effects.

Evidently the result of the Earth colliding with a vast number of cosmic bodies would lead to several outcomes. First the darkened sky would lead to a cooling in the mean temperature of the Earth. Since a reduction in the Earth's mean temperature reduces the amount of liquid water available on the Earth, there would be a rapid variation in sea level. Both of these effects would cause all life to become highly stressed as it was forced to adapt to the new conditions by evolving. Some of the life forms would find no niche to evolve into, and so would simply die out. Other forms would find a new niche and then evolve to fill that position. Whole species of animals would therefore be forced to change or die out.

The Heavy Metal Layer

Although the composition of cosmic matter is similar to the Earth's composition, there are some elements that are more common in cosmic matter. If the Earth has been regularly bombarded with waves of cosmic dust then it would seem logical that some trace of this increase in cosmic material should exist on the Earth. This increase has been observed in unusually high concentrations of heavy metals - these metals are characteristic of meteorites and cosmic dust rather than the Earth's crust.

The material involved is mainly iridium. Increases have been observed at the end of the Cretaceous, 65 million years ago when the dinosaurs became extinct. Smaller increases have also been observed at some of the other major extinctions, notably at the end of the Eocene and Pliocene. These increases in iridium can be accounted for by considering that a cloud of dust and meteoroids hit the Earth over a period of tens of thousands, or perhaps millions, of years.

Until recently the iridium at the end of the Cretaceous was taken to be the result of one major impact as suggested by the father and son team of Luis and Walter Alvarez.[1] But the evidence from over 80 different sites doesn't fit that simple model and the discoverer of the key evidence, Walter Alvarez, has since suggested that a shower of several huge meteoroids or comets may have produced the tell-tale layer over 65 million years ago at the end of the Cretaceous, when many dinosaurs perished.

The Quantity of Material Involved

Although there is evidence that large meteorites from space have repeatedly bombarded the Earth, the amount of material added to the Earth is difficult to quantify. Is this rock a meteorite or has it been carried here after being broken off a mountain by the last Ice Age? These questions are sometimes difficult to answer, and they become even more difficult as the size of the meteorite is reduced. At the size of a grain of dust it becomes virtually impossible to separate the cosmic dust from the normal Earth-blown dust. To overcome this problem scientists usually sample the air high in the stratosphere where the proportion of cosmic dust is high. But as the cosmic dust settles towards the Earth it becomes increasingly contaminated and mixed with other microscopic dust. As the dust finally comes to rest on the Earth's surface it has lost any identity as cosmic dust, and this fact means that no direct evidence is available to estimate how much cosmic dust has covered the Earth in the past.

[1] *Alvarez et al, 1980.*

Chapter 5 - Meteorites and Ice Ages

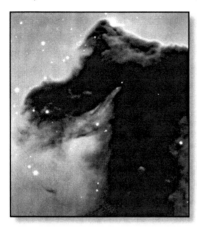

Fig. 5.5 Horsehead Nebula

Cosmic dust can be clearly seen in dark nebula such as the Horsehead Nebula. The Nebula is located to the left of Orion's Belt and is only a small part of a much larger molecular cloud. It is thought that the darkness of the Horsehead Nebula is mainly caused by thick cosmic dust. Stars visible within the centre of the nebula are in the process of forming from this cosmic dust.
© NASA

Despite this, the amount of cosmic material added in the past can be calculated if it is assumed that this cosmic material has accounted for most of the mass added to the Earth. The calculation gives an approximate amount of a million times more cosmic material covering the Earth than the present rate of cosmic bombardment. This value is approximate because the measured amounts of cosmic material entering the Earth's atmosphere tend to vary by hundreds of thousands of tonnes per year. But using this quantity indicates that our solar system should be in a region of space low in interstellar matter, and there should be other regions of space where the density can be seen to be about a million times greater.

Astronomers have observed these cosmic dust clouds in space for many years. One well-known cosmic dust cloud is the Horsehead Nebula as shown in Figure 5.5, and this cloud is so dense that it blocks out the light from stars behind it. The reason we can see dust clouds so clearly is that we are viewing this gigantic mass of dust and small planetoids from a vast distance. By measuring the amount of light that passes through these clouds of interstellar gas and dust an estimation of the density of these clouds has been made by astronomers. Space outside these clouds has been estimated to contain one or two atoms per cubic centimetre, but these clouds are so dense they contain a million times more matter per cubic centimetre.

Even these 'dense' clouds of interstellar matter are very unsubstantial since they approach the best vacuums that can be achieved in a laboratory on Earth, and it is generally thought that only single atoms within the clouds stop most of the light that is blocked by these clouds, so if these atoms collect in small

Dinosaurs and the Expanding Earth

Fig. 5.6 Sombrero Hat Galaxy

Rings of cosmic dust can be clearly seen in the Sombrero Hat Galaxy.
© NASA

grains of dust then they would block less light for a given mass. If these clouds of dust also contain small meteoroids and planetoids the amount of matter contained within them could be vastly more than presently estimated.

Even though astronomers are unable to observe these larger objects visually they have inferred their existence by other means. The method they use is to scan the skies in the infra-red (basically heat) region of the electromagnetic spectrum. Outer space is not consistent in the amount of infra-red it emits and this has been attributed to these dust clouds warming up as they form small meteoroids and planetoids.

Discs of Dust

Some of this dust can be observed much closer to the Sun. A thin layer of fine cosmic dust forms a disc that extends from the equator of the Sun to the outermost planets. Normally this layer is invisible to us, but on fine nights the outermost layers of this dust reflects the sunlight to form a band of light in the night sky. This kind of reflected sunlight is known as the zodiac light since the light appears over the zodiac stars.

The tendency for these discs of dust to form in space occurs on several scales. The ring system of Saturn is on a planetary scale. The disc of dust orbiting the equator of the Sun, to produce the zodiac light, is a much larger size of ring of dust. By far the largest known cosmic dust disc is the type that orbits the billions of suns that make up a galaxy. These rings of dust can be clearly seen in the galaxy of M104, the 'Sombrero Hat' galaxy, as shown in Figure 5.6, and a number of other spiral galaxies, such as those in the constellation of Coma Berenices. Our own galaxy is very similar to these galaxies and this central dust ring can be observed in our own galaxy. We are part of it.

Chapter 5 - Meteorites and Ice Ages

Fig. 5.7 Cosmic Seasons

The path of the Sun and Earth around the Milky Way Galaxy indicates that cosmic bombardment should be highly variable over geological time.

The ability to be able to visualise our own galaxy as a spinning disc of stars, planets and dust provides a reason for the variation of cosmic material hitting the Earth. The whole galaxy is rotating so our Sun completes one revolution of the galaxy in 200 million years. All the other stars, planets and dust travel at slightly different speeds and directions, so it is inevitable that the Sun will carry the Earth into regions of denser cosmic dust over geological time.

As well as travelling in the general direction of this spinning disc of gas and dust, the Sun is moving perpendicular to it. If it continued moving in this direction it would leave the galaxy, but this is not what will happen. As the Sun moves further and further from the main mass of the disc, the gravitational attraction of this mass will slow the speed of the Sun. Then it will eventually stop and begin to fall back towards the disc. As it comes closer to the disc its speed will increase faster and faster, until as it reaches the disc its speed will be great enough to carry it through the disc to the other side. Once on the other side of the mass of the disc the Sun will again begin to decrease its speed, until once again it will fall back towards the main mass of the disc.

This process of cosmic seasons is illustrated in Figure 5.7 and must have been repeating itself since the Earth first began to form. Since the Earth passes across this central dust lane every 30 million years or so, it has been suggested that the rates of extinctions may also occur with similar timing as the Earth bobs up and down on its journey around the galaxy.

Some scientists have noted that an increase in the number of impact craters occurs roughly every 30 million years and these figures coincide with a possible period of mass extinctions. Other activity appears to come in waves - not only the dates of impact craters, but also the cycles in the Earth's volcanic activity.

The amount of matter in the form of suns has been calculated to represent only a small fraction of the total matter in the universe. By observing the motion of the universe, the force of gravitational attraction and hence the required total mass of the universe have been calculated. But these calculations reveal that there must be much more mass than we can see in the visible universe. The evidence indicates that all this hidden mass of dust and meteoroids amounts to a lot of matter and it has been estimated that this missing mass may amount to 90% more mass than that contained in all the visible suns of our universe.

Taking all this evidence together, there is sufficient reason to believe that the Earth may well have passed through a massive cosmic dust cloud, not just once, but many times in its history.

Checking the Rate of the Earth's Expansion

If the Earth has gradually formed from the accumulation of cosmic material then it should be possible to calculate the rate of this build-up, and then compare this build-up with the predicted rate of expansion of the Earth from geological data. The period over which this expansion takes place is so long that the cosmic bombardment can be considered to be a constant process. A period of reduced cosmic bombardment might well last for tens to hundreds of thousands of years, but this time is so short in comparison with the formation of the Earth that these fluctuations can be ignored.

The formation of the Earth can therefore be imagined as the creation of a gravitational field about the Earth in a constant stream of cosmic material. At first the gravitational field is small due to the small size of the Earth and hence the amount of material that is captured by the Earth is small. As the Earth accumulates more mass the size of this gravitational field increases and the Earth increases in size at a faster and faster rate.

The signs of Earth expansion that are present on the ocean floor are only reliable up to about 200 million years ago, but they do seem to show the same rate of increasing Earth expansion. Also, by projecting the expansion of the Earth backwards in time the Earth appears to have begun to form about 4,600 million years ago, which is the current estimate of the beginning of its formation based on the age of rock and meteorite samples.

Chapter 5 - Meteorites and Ice Ages

The Transport of Material into the Core of the Earth

At first I was puzzled by the concept of an Earth that was slowly covered in cosmic material since this would result in the oldest portion of the Earth as the central section with the newer material slowly covering the outer layers. But this is not what has happened. Some of the oldest rocks of the Earth are found on the continents, while the newest rocks are on the ocean floor formed from the molten lava welling up in the centre of the ocean rifts. So how could the idea of the Earth being covered in cosmic material fit these facts?

I needed to explain how layers of cosmic material could be transported into the interior of the Earth. The first part of the story is easy enough to explain. Rain falling through the air quickly removes the dust suspended in the atmosphere. Meteorites of various sizes that become lodged in the ground are also continually eroded by the rainwater.

Earth is one of the most chemically reactive and geologically active planets in the solar system. Above the outer layers of our planet is an atmosphere that contains a highly reactive gas, oxygen - the same gas that gives us life. Oxygen will attempt to react with most elements at low temperatures. Some elements form an oxidised layer on their surface that protects them from further attack. Aluminium and copper are examples of this, both these metals being considered stable in air while any newly-machined surfaces can be seen to rapidly build up a protective oxidised layer that prevents further corrosion. Other elements are less fortunate and form oxidised layers that break away so new surfaces constantly form for the oxygen to attack. These relatively mild reactions all occur at room temperature but at progressively higher temperatures the reactions become more violent. As the temperature rises the reactions begin to emit heat until the reactions become self-sustaining and fire results.

Our planet is covered by water over three quarters of its surface and this water is almost as reactive as the atmosphere. As well as containing dissolved oxygen it can hold several other elements in it. The effect is obvious when we observe how many salts are dissolved in the sea.

Many of the rocks of the world are easily dissolved by water but by far the most soluble of the commonly occurring rocks are the limestones. Some of the best-known cave systems of the world are carved in these solid rocks. Starting as a fracture only

Dinosaurs and the Expanding Earth

a few microns wide, water seeps down and dissolves a wider opening until a flow develops, and as the water moves faster it begins to erode the rock. In time, perhaps in as little as a few thousand years, great cave systems are formed which are large enough to be explored.

One of the largest cave entrances in the world is the Bournillon Cave in the Vercous Mountains near Grenoble, France. It opens at the base of an immense vertical cliff of white limestone, the opening being over 80 metres (262 ft) high and 30 metres (98 ft) wide. Another cave, the Kara River Cave in the East Indies is so tall at 100 metres (328 ft) high and 100 metres (328 ft) wide that it frequently has its own cloud system inside it.

As the water percolates down through the limestone it dissolves the rock, reaching solutions of 300 parts per million. As the water enters the cave the limestone is forced out of solution when it hits the cave atmosphere. The calcium carbonate precipitates out onto the wall of the caves and as water drips from the roof stalactites and stalagmites are formed as shown in Figure 5.8. These can grow to immense sizes; there is one over 15 metres (49 ft) high and wide in the Favot Cave in the French Vercous.

Eventually, the water carries these dissolved salts into small streams and then rivers which feed into the oceans. As the concentrations of the solids and salts build up the metals begin to come out of solution and are deposited as sediments.

Vast quantities of sediments are moved around the Earth by wind and water. Sediment moved by wind is usually very fine dust or sand but because it is so fine it can enter the upper

Fig. 5.8 Stalactites and Stalagmites

Many large cave systems throughout the world have spectacular displays of large limestone formations of stalactites and stalagmites.

Chapter 5 - Meteorites and Ice Ages

atmosphere where it can be transported around the globe. Dust from the Sahara regularly reaches the UK and has been found as far away as the Caribbean.

Large quantities of sediment are moved by water in rivers and floods. Rivers can carry sediment of sand and gravel size but large floods can carry rocks and sometimes boulders.

Coastal erosion removes large quantities of material due to the action of waves and currents. Glaciers also carry large amounts of sediment which may contain boulders up to several metres in diameter. Sediment also creeps down slopes in landslides.

Much of the sediment naturally flows into the ocean and then is finally swept into ocean trenches to supply great volumes of new material.

Some features of the sediment layers in trenches have been interpreted as 'accretionary prisms' of material that have been scraped off the ocean floor due to plate movements on a Constant Diameter Earth. But these layers of sediments can also be interpreted as sedimentary subsidence after first flowing off the continents as sediments and then settling and subsiding down towards the trenches. The fact that these sediments are only found close to continents seems to support this view. If they were 'accretionary prisms' features on a Constant Diameter Earth then they should be visible at all the ocean trenches as the top material is scraped off, but they are only found in trenches close to continents where sediment is washed off the continents.

Isolated Subduction

From the 1920s through to the present day, several researchers used gravimeters to measure the gravity over ocean trenches and found that these were all areas of reduced gravity. This was interpreted as indicating that trenches are areas where material is moving or sliding down into the Earth thereby leaving a hole, or trench, which has reduced gravity.

As the weight of sediment on an ocean trench increases, the floor of the trench begins to slowly sink downward into the Earth's mantle for perhaps tens of thousands of years. If the top layer of the trench continues to be replaced with denser sediments the ocean trench continues its downward motion for millions of years, slowly transporting denser sediments into the interior of the Earth. This is a fairly good description of the ocean trench subduction zones found mainly in the Pacific Ocean which are usually tens of kilometres long.

The subduction of material into the Earth at these ocean trenches has already been used as part of the evidence for a Convection Cell theory to explain the destruction of ocean floor on a Constant Diameter Earth. The use of subduction to explain the Constant Diameter Earth model relies on connecting these subduction zones to the whole ocean floor to produce gigantic Convection Cells spanning the entire ocean floor. But there is no evidence that subduction zones are connected to the whole ocean floor. All subduction seems to be isolated from most of the ocean floor surrounding it.

Almost all the current ideas about what is happening in these ocean trenches are just as relevant to the Expanding Earth as they are to the Convection Cell theory previously shown in Figure 3.6. The major difference is that an ocean trench no longer needs to be connected to a slab of the ocean floor. The ocean trench can be imagined as Isolated Subduction driven by the influx of denser new sediments forcing themselves into the depths of the trench as shown in Figure 5.9.

Since this Isolated Subduction is unconnected to most of the ocean floor it cannot force the convection of the whole ocean floor into the trench. The previous problems in using subduction to explain the Constant Diameter Earth model no longer exist: the driving force for Isolated Subduction is large

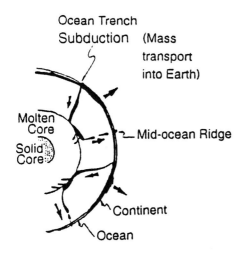

SECTION THROUGH AN EXPANDING EARTH

Fig. 5.9 Mass Transport into Earth

An Increasing Mass Expanding Earth would be very similar, in many details, to the current giant Convection Cell theory of the Constant Diameter Earth as previously illustrated in Figure 3.6. Isolated ocean trench subduction would gradually transfer new denser material into the interior of the Earth over millions of years.

Chapter 5 - Meteorites and Ice Ages

enough since the whole ocean floor does not need to be dragged down into the Earth's interior, and the difference in length between the mid-ocean ridges and the subduction zones is not a problem since they no longer need to be equal lengths.

As this denser material is carried down into the interior of the Earth, the higher temperatures and pressures begin to make the chemical compositions unstable. Eventually the chemicals reform into new compounds. This reformation causes earthquakes at 50 to 100 km (30 to 60 miles) below the surface of the Earth, close to the ocean trench, and it is these earthquakes which have been plotted to outline the flow of new material being subducted into the interior of the Earth. As the elements reform into new compounds, some of the material separates out into heavier and lighter compositions. The heavier compounds continue on their downward journey into the Earth but the lighter compounds begin to force their way back to the surface of the Earth. These lighter compounds eventually emerge to form volcanoes.

Most of the volcanoes of the world are close to the ocean trenches from which they originally began. Part of the material that emerges from these volcanoes are the gases and water which make up the atmosphere and the oceans. Volcanoes emit vast quantities of carbon dioxide, sulphur and water. Thus the constant influx of cosmic material has not only expanded the Earth, it has also allowed the atmosphere to become denser and the oceans to fill the expanding ocean basins.

If the connection between cosmic material, ocean trenches and volcanoes does exist then there should have been major eruptions of lava as the lighter compound forced its way back to the surface a few million years after the Earth was hit by a cosmic storm. These outpourings of great lava flows have been observed in the geological record. One massive lava eruption formed the Deccan Traps in Indian about 65 million years ago and is one of the largest volcanic features of the Earth. The lava flows seem to be multiple layers of lava that formed a flood basalt lava layer that is now 2 km (1.2 mile) thick spread over an area of 500,000 square kilometres (193,000 sq miles). Before this area was eroded down to its modern size it is estimated that the original Deccan Traps lava field was approximately half the size of modern India.

It has already been noted that a correlation exists between the amount of iridium deposited at certain times and the deaths of large-sized species. These iridium variations from the normal appear to coincide with the extinction of the dinosaurs at the close of the Cretaceous, and show a possible correlation with the extinction at the close of the Eocene and a good correlation with the extinction at the close of the Pliocene.

The evidence seems to indicate that iridium deposits are related to cosmic material, massive lava flows and mass extinctions which all occurred at similar times.

The Ice Ages

These waves of cosmic dust and other cosmic material would have more effects. The Sun would be blotted out and the Earth would cool. The Earth would be thrown into an Ice Age with the polar caps becoming greatly increased. After this wave of cosmic material had passed the Earth, the heat of the Sun would return to warm the Earth. The great polar caps would melt and the whole Earth would increase its temperature to return it to its previous warm conditions. On both the long and short timescale, there would be no pattern to these glaciations and warm periods. Some glaciations would last for a short time and others could last for tens of thousands of years. The same is true of the warm periods. This fits well with what is known of the last Ice Age.

The growth of glaciers has often been linked with a global change in the Earth's climate and it is generally estimated that the sea level fell by about 100 metres (328 feet) during the last Ice Age as the glaciers locked up large amounts of water. At certain times in the Earth's history the climate has been too warm for glaciers to form. But ice expansion seems to have been caused by external cooling of the global temperature.

Many theories have been proposed to explain the Ice Ages, but the present most widely accepted is the Milankovitch Model. This theory relates the record of temperature variations of the Earth to match the 100,000 year, 41,000 year and 22,000 year rhythms of the Earth's orbit around the Sun. But the Milankovitch Model places the Earth in the middle of an Ice Age now, and this is obviously not the case. If it is accepted that the normal state of the Earth today should be an Ice Age, why are we presently enjoying an Interglacial?

Besides this problem Ice Ages have always occurred simultaneously in both hemispheres of the globe. The Earth slowly changes the direction of its tilt in space, like a giant version of a child's spinning top as it slows down, and this is

Chapter 5 - Meteorites and Ice Ages

known as precession. The Milankovitch Model suggest that the northern hemisphere Ice Age was caused by the Earth's 26,000 year precession changing the angle of the Earth's tilt towards the Sun. This might explain the northern hemisphere, but why does the southern hemisphere follow the same pattern of Ice Ages? If the theory was correct then the north and south hemispheres should show different patterns.

Because of these problems several people have suggested alternative reasons for the Ice Ages. One suggestion came from Fred Hoyle, the British astrophysicist. He considers the effects of the Milankovitch Model too small to account for temperature fluctuations sufficient to throw the Earth into an Ice Age and therefore argues that the impact of a large stone meteorite would cause an Ice Age.

Other people have also suggested theories related to cosmic causes. As long ago as 1921, H. Shapley was suggesting that the solar system was affected by 'cosmic seasons' in its orbit around the Milky Way galaxy. He continued advocating that galactic rotation of the Sun and planets around the centre of our galaxy caused cosmic seasons for a number of years in various scientific periodicals.[1]

Later, in 1975, W.H. McCrea used more recent astronomical data to make the idea more attractive. Since the galaxy is a flattened disc of stars with two spiral arms, rather like the spiral galaxy M51 shown in Figure 5.10, it was suggested that the Ice Ages resulted when the solar system entered the dust lane of a spiral arm. McCrea related the occurrence of the Ice Ages to the passage of the solar system through these dust lanes. Within these dust lanes the distribution of the dust is patchy with

---

[1] *Shapley, 1949.*

Fig. 5.10 Spiral Galaxy M51
M51 Spiral Galaxy is similar to our own galaxy and the dust lanes can clearly be seen.
© *NASA.*

repeated changes in density. This change affects the solar output from the Sun so McCrea thought that:

> ... there should ... have been an Ice Epoch during the past few million years extending up to near the present time.[1]

This concept fits the known facts. The last great Ice Age lasted about 2,500,000 years. Several times over this period great ice caps spread from the polar regions to form ice sheets that spread into today's temperate latitudes. In the northern hemisphere these sheets covered all Canada and sometimes advanced southwards into the United States of America as far as Long Island and the Ohio and Missouri rivers. They spread onto the Baltic Shield, the West Siberian lowland and half of Great Britain. The glaciers carved great valleys in Scotland's mountains and the American Rockies. They carved grooves in rocks that are now visible in New York's Central Park, and they diverted the course of the Thames tens of thousands of years before London was built. Similar ice sheets spread up to reach the southern continents until a third of the Earth was covered in snow and ice.

During the last 100,000 years the pattern was typical of those that preceded it. The weather was mostly colder than the present temperature, from 100,000 to 50,000 years ago, but mild compared to later conditions. A brief warm spell occurred about 43,000 years ago when temperatures were probably higher than now. Following the period of warm weather the temperatures fell swiftly, then they remained mild until about 25,000 years ago when the climate became severe. The glaciers moved south to extend over a large portion of North America and Europe. This situation lasted until about 15,000 years ago. Then the ice caps melted and warm conditions returned. There was a brief return to arctic conditions about 11,000 years ago but our present warm period has lasted for 10,000 years.

Cosmic Winters

Two authors, Victor Clube (an astrophysicist at the University of Oxford in the UK) and Bill Napier (a well-known science writer and astronomer) have collaborated to publish two books, *The Cosmic Serpent* and *The Cosmic Winter*, which argue that rains of cosmic material visit the Earth from time to time. The most

---
[1] *McCrea, 1975.*

Chapter 5 - Meteorites and Ice Ages

recent cosmic storms have been severe enough to destroy ancient civilisations and plunge mankind into Dark Ages.[1] They uncovered a lost history of celestial catastrophe which reveals that civilisation could well come to an abrupt end in a rain of fire followed by an icy cosmic winter.

In the *Cosmic Serpent* they propose a detailed mechanism to illustrate how the Sun could capture interstellar cosmic material in its obit around the galaxy. The Sun and planets are presently moving away from an active region of young stars, gas and dust known as Gould's Belt and they calculate that the Sun must have passed through this region of star formation only 10 million years ago.

In the first part of their second book, *The Cosmic Winter*, the authors gather various clues from myths and legends around the world that point to ancient cosmic impacts in the past. They suggest that the ancient 'Gods' were probably not based on fiction but are real descriptions of comets from active periods of cosmic bombardment in history. These ancient comets are described as moving around the sky and interfering in worldly affairs with deadly bombardments.

In the second part of this book they summarise the current scientific knowledge about comets and suggest that when a large mostly uninhabited region of central Russia was devastated in 1908, by what is known as the Tunguska impact, only a small fragment of comet Encke was to blame.

They conclude that all the evidence indicates that impacts tend to be concentrated into brief periods with multiple impacts being much more severe than currently supposed.

Small Comets

Some scientists believe they may have more direct evidence of a constant stream of cosmic impacts happening now. Louis A. Frank, a Professor of Physics at the University of Iowa, made startling observations that both he and his colleagues interpreted as a vast number of comet-like objects disintegrating as they hit the upper atmosphere.

In the early 1980s, Frank and his colleagues began making observations of puzzling spots appearing in ultraviolet images of the Earth's upper atmosphere, which had been taken by a satellite called Dynamics Explorer. After carefully examining all

---

[1] *Clube & Napier, 1982 & 1990.*

Dinosaurs and the Expanding Earth

the evidence, the only explanation for this seemed to be that there were about ten million comet-like objects hitting the Earth's outer atmosphere every year, each one the size and weight of a small house.

Frank and his colleagues published their results in two papers in the 1986 issue of *Geophysical Research* which started a controversy when they described their observations of a fairly constant stream of cosmic matter vastly greater than anyone had ever imagined:

> Large, transient decreases of atmospheric dayglow intensities at ultraviolet wavelengths ... are interpreted in terms of an influx of heretofore undetected comet-like objects. ...
>
> The mass of each of these comet-like objects is ~ 10^8 gm, or ~ 100 tons. The global mass accretion rate by the Earth's atmosphere is ~ 10^{12} kg/year.[1]

In his book *The Big Splash*, Frank described how astonishing the results seemed.[2] At first they had thought the spots on their ultraviolet images might be due to faulty instruments but after ruling out every possibility they could think of they had to face the fact that some sort of real event was causing the dark spots.

The likeliest candidate was vast amounts of water. Judging by the diameter of the dark spots it amounted to about 100 tonnes for each spot. The most probable explanation was that relatively loose-packed balls of water-snow comets were breaking up as they hit the atmosphere, until they expanded into a thin ball of mist-like gas some 48 kilometres (30 miles) across.

The remnants of the small comets would continue down into the atmosphere until they slowed to subsonic speeds and finally mixed with the air in the upper atmosphere.

There is nothing controversial in this description; the big question was the number of these objects. The rate of impact was twenty per minute. Every minute, twenty 100 tonne comets about the size of a house were slamming into the Earth's atmosphere. If this rate was constant over the age of the Earth then there was enough water to fill the oceans. This was a startling conclusion.

Why had no one noticed this vast number of small comets? The answer given by Frank is that they are very difficult to

---

[1] *Frank et al, 1986.*
[2] *Frank, 1990.*

Chapter 5 - Meteorites and Ice Ages

observe and once they enter the atmosphere they mix with it and effectively disappear. Man-made comet-like material was sent up by rocket into the upper atmosphere and produced the same dark spot on new ultraviolet images taken by the Dynamics Explorer satellite. This seemed to confirm the small comet explanation.

In 1988 a dedicated search using a telescope found small dark objects in near-Earth space exactly where the ultraviolet images predicted. The rate of detection of these objects was the same rate as the atmospheric holes identified on the ultraviolet images from Dynamic Explorer satellite. It all fitted.

The discovery of the small comets is based on a set of observations at the very limit of present-day technology and not everyone agreed that they indicated small comets. It stretched the limits of comprehension and some sections of the scientific world responded with indignation.

The number of small comets and their size tended to strike a note of panic in people. After the small comets were observed by the telescope a paper describing the results was submitted to *Geophysical Research*. The referees at first decided they could only publish the results if the streaks were identified on two successive images. But when the two successive images were produced the paper was again rejected because this time it was decided that three successive images were required. By this time the review process had been ongoing for half a year. There seemed to be nothing wrong with the paper apart from the fact that people were simply frightened to publish it.

The scientific community in the United States was in such turmoil that it became obvious that papers supporting the concept of small comets had become impossible to publish there. Eventually the results were published in European journals.[1]

Frank and his team believe future research will produce more surprises. For them, this is only the beginning of an exciting phase of discovery.

If their estimate of small comets is correct then there are ten million comet-like objects, each about 100 tonnes, hitting the upper atmosphere each year. The yearly amount of comet-like water hitting the Earth amounts to 1×10^9 tonnes - 1,000,000,000 tonnes per year. It's not the 1.7×10^{13} tonne per year predicted by the Expanding Earth but it is vastly more than the previous common estimates of cosmic material hitting the

---

[1] *Yeates, 1989 and Frank et al, 1990.*

Earth. And since the amount of cosmic material can easily vary over geological time the amount of material could have easily been much greater in the past.

Although the views of these scientists are generally considered to be at the extreme limits of acceptable thinking, I must conclude that they may be only conservative estimates of the rate of cosmic bombardment. The volume of mass increase predicted by the Increasing Mass Expanding Earth quantifies the amount of material from asteroids, comets and cosmic dust which must strike the Earth over geological time. We must imagine a truly vast amount of cosmic bombardment continuing for millions of years.

The explanation of small comets regularly hitting the Earth's upper atmosphere has been significantly restricted by current orthodox thinking. The dark spots on the ultraviolet images are readily explained by the vast amounts of water contained in small comets, but comets are generally considered to be similar to dirty snowballs comprising one-third to two-thirds water with the rest made up from cosmic dust. The critics of Frank's small comet theory asked for evidence of all the extraterrestrial dust that would have fallen to Earth from the small comets in the past, since if the small comets were similar to other comets there would be a vast amount of cosmic dust added to the Earth each year - something in the region of 1,000,000,000 tonnes of dust per year. Frank effectively side-stepped this argument by suggesting the small comets contained so little dust that they were almost pure water. But I believe future research into these small comets will reveal that they do contain large amounts of cosmic dust. This will have a major impact on the highly controversial notion that cosmic dust increases the mass of the Earth over geological time to produce an Increasing Mass Expanding Earth.

Linking Mass Extinctions to Climate

Cosmic bombardment has been related to mass extinctions through the various effects of extreme climate change by some geologists. One example of the close connection between climate and life is the world-wide reduction in ocean life which coincided with the reduction in water level as it became frozen at the poles during the Ice Age. The marine animals that occupied the large areas of the shallow sea floor were the most severely affected but many extinctions also affected deep ocean life at the same time.

Chapter 5 - Meteorites and Ice Ages

Climatic change can eliminate vast numbers of species with relative ease and can operate on a global scale because its effects touch all animals. Life must make rapid changes to adjust to an extreme variation in climate or become extinct.

If cosmic bombardment, global climate and mass extinctions are related, it should be possible to observe an increase in mass extinctions as cosmic bombardment increased, and several palaeontologists believe they have found close correlations. Estimates vary by a few million years, but mass extinctions appear to occur at intervals of between 28-32 million years. These estimates of the frequency of mass extinctions mean there should be, or have been, a mass extinction within a few million years from now.

This theory seems to fit the facts since there have been many species of animal that have become extinct within the last few million years. The more famous are the mammoth, the sabre-tooth cat and the cave bear. These animals became extinct in Europe and North America during similar time periods and are well known.

While these extinctions were occurring in the northern continents, other animals were suffering a similar fate in the other continents of the world. In South America the giant relatives of the sloth, armadillo and anteater all disappeared. The same fate befell the giant kangaroo and platypus of Australia.

All over the world many species disappeared. These extinctions are only those we know from the few fossils we have discovered. Presumably there must be many more species which became extinct at these times and left no trace of their existence.

Little Ice Ages

Cold periods in the Earth's weather have continued to occur after the end of the last Ice Age. Fortunately for us, these temperature variations have been much less extreme than during the Ice Age and only produced cold periods lasting from a few decades to centuries instead of the tens to hundreds of thousands of years of a true Ice Age. Because of their reduced severity they are called Little Ice Ages.

Some periods in recorded history were slightly warmer than today. In the last century B.C. the Romans were cultivating olive and vine plants in regions where plants cannot grow today due to cold weather. These crops began to fail as a Little Ice Age introduced intense periods of cold. Further evidence of the cooling climate is provided by the discovery of Roman gold mines high up above the present snow line in the Austrian Alps.

By the fifth century A.D. severe cold conditions had affected the world. It has even been suggested that this cooling event may have been a major factor in the fall of Rome since a deteriorating climate would have caused difficulties in their northern territories.

After this colder period of weather, the next warm interval in about 1,000 A.D. offered much improved conditions. The warmth in Europe was provided by mild, wet winters to feed the land and plants. During this time the Norse expanded to settle in Iceland and Greenland and during the last decade of the tenth century the famous Norse leader Leif Eriksson is believed to have taken advantage of the warm weather to found a colony on the North American mainland.

Since then there have been occasions when the same cold conditions that produced the Ice Ages threatened to return with Little Ice Ages lowering the temperature of the whole world. The Little Ice Age winter weather has regularly become so extreme that the River Thames in London was completely frozen on several occasions. This freezing of the River Thames is unthinkable in our present warm period, so it provides a clear guide to the severity of the Little Ice Ages. One of the most severe winters in 1269-70 completely froze the Thames downstream for months. Between 1407 and 1565 the river froze over six times.

Both Henry VIII and Elizabeth I travelled across the frozen surface of the river and the ice was used to hold several 'Frost Fairs' with tents, bowling, shooting and dancing on the ice, as recorded by engravings such as that shown in Figure 5.11. The ice became so thick that great fires were built and parties of many people gathered on the river. Later, in the seventeenth century, the river froze over ten times. Finally, at the end of the last Little Ice Age the warming trend meant that the river has never again frozen completely.

The effects of the Little Ice Age were not restricted to London, and records across the whole of Europe tell of a similar pattern, with many lakes and rivers freezing over. The records of these times from other parts of the world are less complete, but it seems that several changes occurred which fit into the pattern of the Little Ice Age.

With the onset of the Little Ice Age, the cold weather resulted in the collapse of the Greenland colony by the fourteenth

Chapter 5 - Meteorites and Ice Ages

Fig. 5.11 Frost Fairs

During the Little Ice Age the weather became so cold many rivers throughout the world completely froze in winter. During the worst recorded frost, Great Frost Fairs were held on the solid ice as depicted on the engraving of a London Frost Fair held on the river Thames during the reign of King Charles II in 1683.

century and with the breaking of contact with their homelands, all trace of any surviving Norse in North America was lost. The only remaining colony, in Iceland, survived through its value as a whaling base, as the cold gripped much of Europe.

Little Ice Ages were not only periods of intense cold, but also periods of dry weather. In North Africa the periods of the Little Ice Ages brought dry conditions, especially on the fringes of the Sahara. Many summers in Europe during the second half of the seventeenth century were hot and dry with the threat of drought always present. Historically, both the Plague and the Great Fire

Dinosaurs and the Expanding Earth

of London are associated with the droughts that occurred during the Little Ice Age.

The weather records of China and Japan show the same pattern of warm and colder weather. Chinese records extend back to the Chou Dynasty in 1066 BC and give detailed records of great droughts and freezing winters. These records confirm the modern interpretation of the European Little Ice Ages.

These periods of severe weather brought many hardships for the people. In the English county of East Anglia whole villages ceased to exist as the repeated failure of crops brought starvation followed by the Plague spreading out to kill the weakened survivors. Modern historians, who have scrutinised the records of these times, have described how the local populations gradually dwindled.

A Message for the Future

The world is presently becoming warmer. Most climatologists predict an overall rise in temperature of about 3°C in the next 50 years. This increase will make the world warmer than it has been for 100,000 years.

But on a geological timescale it seems certain that another Ice Age will return and history tells us that cold weather is much more difficult to adapt to than warm weather. All the evidence of repeated Ice Ages seems to be linked to mass extinctions and massive lava flows. This picture of events fits in well with the idea of waves of cosmic dust, and other cosmic material, drifting past the Sun and planets and would indicate that the amount of cosmic material hitting the Earth is very low at the present time. It also gives a hope of predicting when the next Ice Age is about to occur. By mapping the amount and speed of the cosmic material that is presently approaching our solar system we could estimate the time and severity of it.

Astronomers can see the process of star and planet formation happening now. Since the Sun is orbiting the galactic centre, over geological time the Earth could easily have moved into one of these planet formation regions. Over geological time the Earth would have been continually bombarded by asteroids, comets and other cosmic material. The common view is that the amount of cosmic material is too small to account for the mass of an Expanding Earth, but the evidence tells a different story. Thousands of large near-Earth objects have been tracked which might hit the Earth at any time and scientists like Clube and

Chapter 5 - Meteorites and Ice Ages

Napier have presented evidence that some large impacts have destroyed ancient civilisations. Over geological time there must have been many large impacts.

Other scientists like Frank have presented evidence that a steady stream of comet-like objects are continuing to rain down on the upper atmosphere in amounts vastly more than anyone had ever estimated.

Once this cosmic material has been added to the Earth, it is transported with common geological processes such as erosion, sediment transport and subduction into the interior of the Earth. Separation within the Earth's core then occurs.

All these facts point towards the startling conclusion that, on a geological timescale, the Earth has been bombarded by vast amounts of cosmic material sufficient to increase the mass and diameter of the Earth - an Increasing Mass Expanding Earth.

6 - The Solar System

The concept of an Earth which has been expanding while life has evolved on it has significant effects on how we must view the whole universe. If the Earth has formed by expanding in size then it is likely that the other planets in our solar system have also formed in the same manner. It must be reasonable to assume that if the planets in our own solar system seem to be forming in this manner, then all the planets in the universe should be forming in exactly the same fashion.

Every planet will go through the same process. A few specks of dust will begin to clamp together to form a meteorite. From a small irregular-shaped meteorite it will become large enough for its gravity to form it into a spherical shape. At first it will be too small to hold any gases at its surface, but after eons of time it will become large enough to hold an atmosphere.

A planet's atmosphere forms in a roundabout route. As a planet increases in size internal activity begins to produce volcanoes on the surface. As well as magma, these volcanoes emit large amounts of gas and water vapour and it is this gas which develops into the atmosphere. As the atmosphere increases in density it begins the process of weathering with wind storms changing the whole surface of the planet by eroding old meteor craters.

Earth is the largest of the inner planets and moons in our own solar system and one of the most significant features of the Earth is that it is being split apart at the mid-oceanic ridges. These extend around the whole Earth and the only reason we did not notice them until the 1950s is that they are hidden under kilometres of water.

These mid-ocean ridges are large-scale expansion features of the Expanding Earth. If expansion is occurring on other moons and planets they should also have similar expansion features.

Chapter 6 - The Solar System

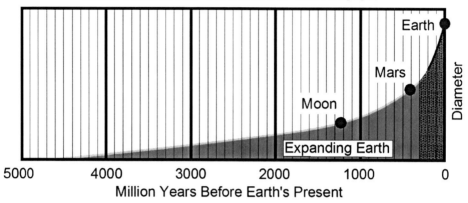

Fig. 6.1 Comparison of Planetary Expansion

All planets should be expanding in a similar manner. If the ancient Earth was once the same diameter as the Moon and Mars the conditions on the ancient Earth should also be similar. These conditions can be compared to see how well the Expanding Earth theory fits the evidence.

It should be possible to see the results of this expansion with examples of different stages of expansion from the moons and inner planets, starting with the smallest moon, through to the largest of the inner planets - Earth. These expansion features are clearly visible to anyone who looks.

Because planetary expansion is an exponential process, signs of expansion should be small on small bodies like a moon, but greater on larger planet-sized bodies. Once again this is clearly visible.

Since all planets develop in a similar manner it is useful to compare the stage of expansion of the Moon and Mars with the Earth's development, as shown in Figure 6.1. Development of these celestial bodies will be discussed in detail in this chapter but the ancient Earth's development will be discussed in more detail in Chapter 7 when this diagram will be referred to once again.

The Moon

On a clear night it is easy to see darker and lighter patches on the Moon and these patches are the light-coloured highland areas and dark-coloured low-lying areas. From Earth the low-lying areas appear to be some form of liquid that has flowed into the craters and low-lying areas of the Moon - hence they are called the Latin for sea - Maria.

Many straight lines known as rills can also be seen on the lunar surface. The largest is the Sissalis rill, which is 450 km (279 miles) long and has an average width of about 3.5 km (2.2 miles). Smaller rills are found down to minute ones, hardly visible. They occur on all types of terrain and appear to be cracks on the surface of the Moon. Single fault systems, in which the ground on one side of the fault drops relative to the other, are also known. The most famous of these is the Straight Wall which is 115 km (71 miles) long and about 400 m (1312 ft) high.

In his book *The Expanding Earth*, Professor of Physics Pascual Jordan also pointed out that the surface of the Moon had numerous rifts and cracks which seem to have expanded. Some are very wide, like the renowned Valley of the Alps which has the same width along its whole length, as can be clearly seen in Figure 6.2.[1] The Valley of the Alps has the appearance of a rupture rift created during movements in the Moon's crust. A small rill runs down the middle of the valley and the contours on opposite sides of the valley have similar outlines at several places. Jordan therefore concluded that the valley arose from a long crack which broadened out. The valley was then filled in to produce a flat valley bottom, and then new small cracks formed along the centre line on this flat valley floor as the crack continued to expand.

---

[1] *Jordan, 1971.*

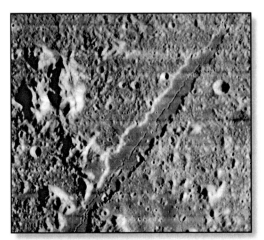

Fig. 6.2 Valley of the Alps

The Valley of the Alps is a wide crack on the surface of the Moon which has been interpreted as a sign of expansion by Professor Jordan.
© *NASA*

Chapter 6 - The Solar System

Most of the Moon's rills are narrower and Jordan placed the Valley of the Alps:

> ... in a somewhat loose morphological group with the Rheita Valley, the Byrgins Rill, the Sirsalis Rill, the Hyginus Rill, the Ariadaeus Rill [see Figure 6.3] and the Schroeter Valley.[1]

Jordan concluded that all these rills resulted from a small expansion of the Moon. A small expansion, very much smaller than the expansion of the Earth, is predicted from the Expanding Earth theory.

The accretion of cosmic dust from outer space is readily demonstrated by photographs of Apollo astronauts' footprints on the Moon, one of which is shown in Figure 6.4. These clearly show how the Moon's surface is an extremely fine, powdery dust. Since there is no weathering on the Moon the most obvious source of this dust is outer space. The Moon is forming from vast amounts of cosmic dust and meteorites.

It is also possible to see numerous craters of all sizes on the Moon and it is now generally agreed that most of the craters have been formed by impacting bodies. The body is likely to be travelling up to 20 km per second (12 miles per sec) as it strikes the Moon's surface and begins to compress it downwards. Huge shock waves develop, and material is squeezed out sideways from between the impacting bodies and the Moon's surface at high speeds. This material has such a high velocity that some

---

[1] Jordan, 1971.

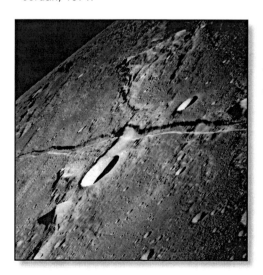

Fig. 6.3 Ariadaeus Rill

Ariadaeus Rill on the Moon as photographed from Apollo 10.
© NASA

can escape the gravity of the Moon and enter space. Meanwhile, the impacting body continues downwards, to form the bowl shape of a new crater. Much of the energy of impact has now been dissipated as heat and the material which is still being ejected begins to fall back to the surface of the Moon. The total time for the average crater to reach this stage would be less than one minute. In the last moments of the formation of the crater, there will be quantities of broken rock and dust which get trapped inside the crater. In larger craters more than 15 km (9 miles) across a central hill sometimes forms.

The highland regions of the Moon are covered with impact craters so densely packed that they form a continuously cratered surface. All ages and size of craters can be seen, ranging from fresh craters with sharply defined rims to older craters which have been filled by some free-flowing material, to the oldest craters where only a few segments of the rim remain visible through the cover of material and recent meteoroids.

The most popular explanation for the 'seas' has been that lava once formed on the Moon and flowed into the old craters and low-lying areas. In some of the older craters it is possible to observe that the rounded floor of the newly-formed crater has been gradually filled to form a flat plateau within the crater walls, and it is common for this material to be the same dark material as the Maria. But when lit from a low angle, the apparently flat Maria show a pattern of thousands of small craters.

The vast numbers of craters on the Moon are a problem for the theory of a Constant Diameter Earth. If the craters occurred geologically recently it would be clear that the Moon was still being created. If this geologically recent accretion happened on the Moon then the same type of accretion would also have

Fig. 6.4 Moon Footprint
This footprint image has become one of the enduring symbols of the Moon visits. Apollo 11 astronaut Edwin Aldrin photographed this footprint in the lunar dust to study its nature. The dust compacted easily leaving a shallow but clear impression of the boot, characteristic of a very fine, dry material.
© NASA

Chapter 6 - The Solar System

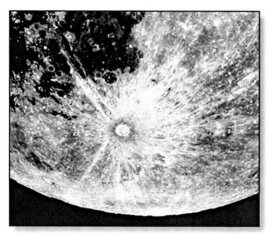

Fig. 6.5 Tycho Crater
The Moon's Tycho Crater appears to be a geologically young crater.
© NASA

happened on the Earth. This reasoning disproves the theory of the Constant Diameter Earth, so in order to avoid this conclusion it is widely assumed that all the craters must be very ancient. The formation of the Moon can then be placed in the distant past and most estimates assume the craters are up to four billion years old. But is this a reasonable estimate of their age?

Many of the Moon's craters appear to be recently formed on a geological timescale. Could these young craters indicate that the Moon is still being formed, in exactly the same manner as the Earth, as discussed in the previous chapter? It would certainly agree with the concept of an Ongoing Cosmic Accretion hypothesis as discussed in Chapter 5.

These young craters are distinguished by their brightness and a brilliant system of light-coloured rays which can extend for up to several hundreds of kilometres from the impact site. One crater which has clearly formed geologically recently is Tycho crater which can be clearly seen with binoculars on a full Moon. The young crater is about 65 km (40 miles) across with a ray system that extends over much of the visible side of the Moon, as seen in Figure 6.5. Over time these bright ray systems will become duller with age, just like the many other craters on the Moon.

One of the more astonishing predictions that must be inferred from the size of the Moon compared to the ancient Earth is that there is a high probability that single cell life is present on the Moon in the form of *Stromatolite* mounds. This is predicted from the fact that when the Earth was a similar size to the Moon there were cones of micro-organisms that formed rock-like mounds, similar to those shown in Figure 7.2 but obviously without the liquid water.

Since there is no liquid water on the Moon, the most likely area where these *Stromatolite* mounds would be found are areas where water-ice is likely to form, probably at the Moon's poles or deep in craters, but they would be very difficult to recognise since their rock-like form would be covered in Moon dust.

Significant amounts of water-ice have been found on the Moon. Most of the Moon is drier than any desert on Earth but researchers have speculated that some areas in permanent shadow might have significant amounts of water-ice. In 2009 NASA deliberately smashed a rocket into a large crater at the lunar south pole and found large quantities of water-ice and water vapour. The experiment used a Centaur rocket stage which impacted into the 100 km (62 miles) wide Cabeus crater which is in permanent shadow. Analysis of the debris that was thrown up from the impact detected even more water-ice and water vapour than scientists had predicted. It is these sorts of areas where single-celled life could be found.

Mars

Mars is a smaller planet than Earth and has many aspects that are remarkably similar to those present on the ancient smaller diameter Earth.

Mars has a nearly continuous 'continental shell' covering the whole of its surface that is only just beginning to rupture to form its own continents. One of the largest geological features on the surface of Mars is a gigantic tear in the crust covering more than a quarter of the circumference of Mars, as can be clearly seen in Figure 6.6. This is Valles Marineris, a gigantic rift valley system that is more than 4,000 km (2,485 miles) long, 200 km (124 miles) wide and up to 7 km (4 miles) deep. This must look almost identical to the smaller ancient Earth when the first gigantic tear in its continental shell first began to form the Pacific.

The formation of such a gigantic crack in the surface of Mars is predicted by the Expanding Earth assuming all planets are expanding. Because of the relatively small size of Mars this tear is comparatively small at the moment and will become bigger in the next 100 million years or so.

The Earth was at this stage of development about 1,000 million years ago. Other conditions on Mars are remarkable for

Chapter 6 - The Solar System

being similar to the conditions on the ancient Earth. Since Mars is further from the Sun than Earth its mean surface temperature is lower. Astrophysicists predict that the Sun has increased its solar output by about 30% during the same time, so the surface temperature on Mars must be close to the temperature that existed on the Earth in Precambrian times.

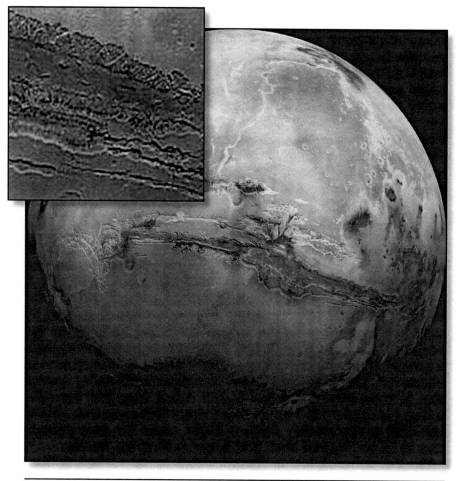

Fig. 6.6 Valles Marineris on Mars

Valles Marineris cuts a wide gash across the face of Mars and such a gigantic crack is predicted by the Expanding Earth theory. The detail shows an enlarged section of the Valles Marineris. Compare the small gashes in the detail with the Moon's rills shown in Figures 6.2 & 6.3.
© *NASA*

The size of Mars resembles the Precambrian Earth, and the sizes of both our Precambrian Earth and present-day Mars limit the development of their atmospheres. Mars appears to be transforming from a small world with no atmosphere into a larger world with a thin atmosphere. Its geology is clearly divided between the old and the new. The old portions of Mars show many impact craters which were formed when the world had no atmosphere. The new portions show the results of a thin atmosphere, generated from volcanic activity, which heralds the start of a new era in the development of the planet.

Although the Martian atmosphere is only a hundredth of the pressure on Earth it still carries clouds of fine yellow dust and white clouds of minute water crystals of ice. The first space missions showed the southern hemisphere of Mars to be covered with many ancient impact craters. But if the Martian atmosphere had existed for a long period, the ancient impact-scarred surface would have been eroded long ago. It seems likely that the atmosphere of Mars has only recently formed on a geological timescale.

Different types of impact craters also indicate that the atmosphere of Mars is fairly new in geological time. Some of the impact craters found on Mars are strikingly different from the majority and these differences show that conditions have changed a great deal. On the edge of the Kassei Valles on the western side of Mars is an impact crater with a diameter of 18 km (11 miles) and many similar craters are found elsewhere on Mars. They have well-defined continuous radial grooves which must have been formed as material was thrown out from the meteoritic impact. Since they are similar to the craters found on the airless and dry surface of the Moon it seems Mars was airless at the time the crater formed.

Fig. 6.7 Arandas Crater

The Arandas crater on Mars appears to show material that has flowed away from the main impact. Is this caused by permafrost lying just below the Martian surface?
© *NASA*

Chapter 6 - The Solar System

Other craters on Mars show a different form. One such is Arandas crater, a 28 km (17 mile) diameter crater in Acidalium Planeta, as seen in Figure 6.7. The material pushed out from this type of crater shows how the material flowed away from the main impact. This flow pattern may well have been caused by permafrost lying just under the Martian surface. As the impact crater was being formed this permafrost would melt, so allowing the material to flow in waves which then became frozen.

The largest known depression, or basin, in the solar system exists on Mars. Known as Hellas Planitia, this depression is 1,600 km (994 miles) across and 5 km (3 miles) deep. Its roughly circular shape suggests that it is an impact crater - one of the largest visible impact craters known in the solar system. The interior floor is some 2-4 km (1-2.5 miles) below the general level of the Martian surface. Hellas Planitia is sometimes mistaken as the south polar cap because it is often one of the brighter regions of the planet and it has been suggested that this may be due to low-lying clouds or frost. Since the atmospheric pressure at the base of this depression would be greater than the average pressure this area would be one of the first to form ice - a complete reversal of what happens on Earth, in fact. This is again due to the fundamental property of water that exists at these low pressures. The conditions present on Mars mean that at lower altitudes the higher pressures allow water-ice to form. The pressure is so low that this water can only exist as ice or vapour since liquid water would boil away.

Amongst these craters are vast areas of collapsed and jumbled terrain and these areas show the beginning of internal activity within the planet. The northern hemisphere of Mars has fewer craters and more volcanic activity. The largest known volcano in the solar system exists on Mars. Olympus Mons is over 20 km (12 miles) high with a diameter at its base of about 600 km (373 miles). Its height is roughly twice that of the largest volcano on Earth - the Hawaiian volcano Mauna Loa, but as this stands on the sea floor it is not nearly as impressive. At the top of Olympus Mons is an area 80 km (50 miles) across which was formed as the top portion of the volcano collapsed back into the central tube where the lava had risen up.

It is noteworthy that the size of the highest volcanoes on Earth are about half the size of the highest Martian volcanoes and gravity on Mars is about half the gravity of Earth. It almost seems that both volcanoes have reached their maximum size for

Fig. 6.8 Martian Surface
NASA's two Viking landers alighted on Mars in 1976 and sent back detailed images of the Martian surface. The view from Viking 1 showed a field of boulders with one dusty foot pad of the spacecraft just visible at the lower right.
© NASA

their own planet's gravity - another case of size limitation under a gravitational field?

Other young features are present on Mars, including large fault systems, wind-formed deposits and small plains. The largest impact craters have nearly all been eroded.

Once again, comparisons of the size of Mars with the ancient Earth indicates there is a high probability that single-celled *Stromatolite* mounds are present where there is water-ice on Mars. These would be difficult to identify since they would be rock-like mounds covered in Martian dust, as shown in Figure 7.2 but without the water. A comparison of these *Stromatolite* mounds with rocks photographed on the surface of Mars, as seen in Figure 6.8, illustrates how difficult it might be to locate these on Mars.

If we look at the graph in Figure 6.1 comparing the estimated expansion of the Earth with the size of Mars positioned on the expansion line, it seems that Mars is developing an atmosphere slower than the ancient Earth. By 400 million years ago the Earth's atmosphere had been stable enough for the previous 200 million years or so to allow seas to develop and multi-cellular life to evolve in them. Or has the Earth developed an atmosphere faster than normal - perhaps because it had a relatively large moon causing more gas to be emitted from the interior of the Earth? In any event, Mars is only within about ±50% of the predicted planetary development stage compared to the ancient Earth. Perhaps this is a normal variation.

Chapter 6 - The Solar System

One interesting concept for the ancient Earth is a Snowball Earth, when great sheets of ice covered most of the Earth sealing the liquid water beneath its surface. The evidence for this, and its implications for the Expanding Earth, will be discussed in the next chapter but could a similar situation have happened on Mars? Could great sheets of ice have covered most of Mars at the same time as the Earth was thrown into an Ice Age?

A Snowball Mars would become almost entirely frozen by forming an ice-covered surface over nearly the whole planet. The Snowball Mars periods would leave evidence for extensive ice and water flows on Mars but these intense cold periods would be punctuated by warm periods comparable with the interglacial warm period we are presently enjoying on Earth.

Because Mars is a small planet, any minor change in global temperature would rapidly affect the whole of Mars. A smaller temperature difference between the poles and equator is natural for a smaller planet like Mars, so a Snowball Mars could rapidly develop an ice-covered surface which could just as rapidly disappear with relatively minor global temperature variations. It seems likely that Mars is now the warmest it has been within the last hundred thousand of years, so only small remnants of this once great ice sheet remain today.

The recent explorations of Mars have provided substantial evidence for major water flows at some time in the past. A Snowball Mars could have produced water sealed under the ice very recently on a geological timescale. It is likely much of Mars was still frozen as recently as a few hundred thousand years ago. As Mars became warmer the glaciers would begin to melt. Because of the low atmospheric pressure any meltwater formed would rapidly evaporate once it was completely exposed to the Martian atmosphere. The low gravity would allow most of this water vapour to rapidly leak into space to produce the dry conditions we presently see on Mars.

When the ancient Earth was a similar size to Mars, multi-cellular life had evolved in the sea but not on land. If liquid water was present on a Snowball Mars, trapped under the frozen ice sheet perhaps a few million years ago, it would provide the conditions necessary to allow multi-cellular life to evolve in that water. Then, as the ice sheets retreated to the poles, the ice-covered seas would disappear and multi-cellular life would be forced to follow the water to the poles.

If we take this concept of a Snowball Mars to its logical conclusion then it could well be that multi-cellular life is still

Dinosaurs and the Expanding Earth

struggling to survive in liquid water beneath the frozen polar ice caps and any remaining isolated glaciers.

In 2009 NASA announced that they had detected substantial amounts of methane in the atmosphere of Mars. Methane is quickly destroyed in the Martian atmosphere so the discovery of sizeable plumes of methane in the northern atmosphere indicates that some ongoing process is releasing the gas. Life produces much of the Earth's methane as it digests food, so one possibility is that this gas is coming from life.

Millions of years in the future liquid water under the frozen ice sheets could once again start to cover Mars and life will again follow. In 500 million years' time, if any future Martians look back at the fossil record of their planet, it would be unlikely that they would notice a few million years when the first multi-cellular life was forced to retreat to the frozen poles, just as we find it difficult to identify those conditions in the ancient fossils of the Earth. Today's Mars seems to be a present-day likeness to the ancient Earth before life evolved in abundance.

Venus

Venus is our sister planet, being only a little smaller than Earth in diameter. The surface of the planet is hidden under a dense atmosphere, but with the use of large radio telescopes and satellite radar, scientists have been able to map the surface of Venus in great detail. They describe how the surface seems to be covered with large volcanoes, trenches and rift valleys similar to the Earth's ocean floor. Artemis Coronia for example is a long deep valley on Venus that is similar in size and shape to a typical ocean trench on Earth.

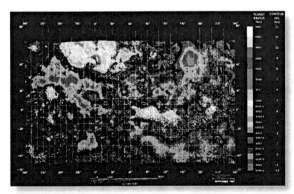

Fig. 6.9 Map of Venus

Map of Venus showing the elevated 'continents' as lighter areas. These differences are even clearer on the original NASA colour map.
© NASA

185

Chapter 6 - The Solar System

There are regions of high ground which may be similar to our continents and these two different areas have been mapped by NASA as shown in Figure 6.9. In the low-lying regions, which on Earth are covered by the ocean, they found flat plains of lava which contained dozens of small vents formed by volcanic cones. The radar also shows that there are relatively few impact craters on the low-lying plains compared to the large highland regions mapped by Soviet spacecraft. This suggests that the highland regions are much older than the low-lying areas. The highland regions would therefore be large old continental-type regions which have broken up as the newer material formed the low-lying areas.

As the temperature on the surface of Venus is 460°C (860°F), life there is considered to be unfeasible.

The Giant Planets

The inner planets and moons display a progression in size up to the Earth. But what happens when a planet becomes larger? Its diameter, mass and surface gravity will all increase. Its atmosphere will become denser at first because this atmosphere will be composed of the heavier gases - just like the minor planets. But as the gravity at the planet's surface increases with its increasing size, the planet's atmosphere will begin to trap the lighter gases of helium and hydrogen. If the planet is cold enough, the immense pressures at the surface of the planet eventually become so great that helium and hydrogen form a solid layer around the planet's surface.

Neptune and Uranus are about four times the diameter of the Earth with an atmosphere of liquid hydrogen. Although their mass has increased, their density has reduced, probably due to the low density of the liquid hydrogen which forms a large volume of the planets. The largest of the solar system planets, Jupiter and Saturn, are over 10 times the diameter of the Earth and their density is also low.

From Brown Dwarfs to Stars

By the time a planet has reached the size of Jupiter, it is attracting large quantities of hydrogen and helium to increase its mass. As this process continues the planet will begin to warm up by thermonuclear fusion and the accretion of new matter, so eventually it will become large enough to become a star. This planet will have become so large that it will begin to emit heat at this phase of its development. At this stage it may be 10 to 20 times the size of Jupiter and it becomes an object known as a 'brown dwarf' which is on the verge of becoming a star. As it grows even larger it eventually produces so much light and heat that it becomes a star.

Distant Galaxies

When astronomers view the night sky they observe an unbelievably vast universe. The distances involved are so great that the speed of light is used as a measurement. We know that a ray of light does not travel instantaneously but in fact moves at 300,000 km in one second (186,000 miles per sec). Even at this vast speed the distances between the stars are so immense that the light travels for great periods of time. The standard unit of measurement of these distances between the stars and galaxies therefore becomes the length of time that a ray of light travels in one year - a light year.

At the extreme limits of the universe lie the quasars. These objects are thought to be the oldest and most distant objects known. They are so far away that their light takes as much as several thousand million years to reach us - it has been travelling since before the Earth began to form.

The furthest galaxies provide the astronomer with evidence that the whole universe was created 20,000 million years ago in a shattering explosion. The heart of this theory lies in a discovery, made in 1913 by the American astronomer Vesto Melvin Slipher, that most galaxies are travelling away from the Earth at very high speeds. The fact that all the galaxies were moving away from the Earth was a surprise. Why should our planet be at the centre of an expanding universe? If Slipher's measurements were correct, the entire universe was moving away from just one point in space. Why we appear to be at the centre of this expansion is still an unanswered question.

During the following decade, what was then the largest telescope in the world was used by Milton Humason and Edwin Hubble to measure the speeds and distances of even fainter galaxies. They found that all the distant galaxies were moving away at high speeds. The most distant was moving at over 1,450 million km per hour (900 million miles per hr).

Chapter 6 - The Solar System

The implications of a universe that is now expanding are that at some time in the past all the matter in the universe must have been packed into one small point in space. Then this matter began to expand at tremendous velocities and with enormous high temperatures. Eventually the matter cooled enough for the first stars to begin to form. By comparing the speed of the galaxies with their distances, the time when all the galaxies were packed together can be calculated. This calculation gives a probable age of the universe of 20,000 million years.

Such an immense amount of time as 20,000 million years is difficult to grasp. The galaxies are thought to have begun to form about 12,000 million years ago. The Earth is thought to be about 4,600 million years old. For most of its life it appeared barren of life apart from microbes deep in its rocks. Then about 600 million years ago the first multi-cellular animals appeared, as worm-like and jellyfish-like creatures. After 200 million years of living in the sea, the first fish evolved into creatures that crawled out of the water; a further 200 million years introduced the dinosaurs, only for them to disappear within 150 million years; the first man-like creature appeared about 3 million years ago, and as the last Ice Age retreated about 10,000 years ago, man took advantage of the change in climate to progress. It's an immense amount of time.

The light from these distant objects shows many dark 'absorption lines' which astronomers think are caused by the light being stopped at one particular frequency by massive clouds of gas and dust. It is believed that these massive clouds collected in a rotating cloud, then as this gas and dust began colliding with others, the kinetic energy turned into heat to form a spinning disc of gas and dust. This gas and dust provided the raw material for the formation of the stars of the galaxies. We have already seen in the previous chapter how these dense clouds of cosmic dust, such as the Horsehead Nebula in Figure 5.5, are still in the process of forming suns and planets from this dust.

By comparison with the vast distances between the galaxies, the size of most galaxies is small, but it would still take light about 100,000 years to cross from end to end in our own galaxy. There are many galaxies like our own - spiral galaxies of a spinning disc of stars like the Sun, as well as substantial quantities of interstellar dust and gas. In most spiral galaxies the disc is very thin in comparison to its diameter, as this dust and gas collapse to form the raw material of new stars. The variation in the brightness of the spiral arms show us where most stars are forming. In the centre of the spiral disc of spinning stars is an oval of ancient stars known as the bulge. In some galaxies the central bulge is tiny, and in others it is huge, almost obscuring the outer disc of stars.

If the Earth had been a constant size for eons of time then the other stars and universes should be static and unchanging. This is not the case, for astronomers describe galaxies which are continuing to form now, with the oldest suns in the centre and the newer suns towards the outer rim. The oldest suns of a galaxy, the red giants, are typically clustered around the central core. Further out in the arms of the galaxies the stars become progressively younger and bluer with youth. Intermingled among these stars are clouds of dust and gas from which the stars and planets are being formed at this very moment in time. A wealth of information tells the story of how our Sun and galaxy was formed, and how our Sun will become a red giant and then die.

Milky Way Galaxy

Our own galaxy, the Milky Way, is a typical spiral galaxy composed of thousands of millions of stars, planets, dust and gas, spinning in a flattened disc around the central bulge. The speed of rotation is about 200 million years, so when the Sun was on this side of the galaxy it was the beginning of the reign of the dinosaurs. When astronomers look at how fast individual galaxies rotate about their own axes, and about each other, they find that the rotation speeds are too great for the amount of visible matter they contain. They should fly apart. The only reasonable explanation is that they contain a lot of hidden or dark matter, providing additional gravitational force to hold the spinning disc of stars together.

It is only recently that the vast amounts of this dark matter have become clear to astronomers. As noted in the science journal *Nature,* one galaxy in particular, MG1654+1346, has a mass of between eight to sixteen times more than the total mass of the visible stars it contains.[1] This galaxy has a quasar aligned almost perfectly behind it and this mass causes the radio waves from the quasar to be bent by the gravitational field of the galaxy to form a distorted arc-like image around the galaxy. These

---

[1] *Langston et al, 1990.*

Chapter 6 - The Solar System

images are known as an Einstein Ring. Because this particular galaxy forms an almost perfect lens, the mass of the galaxy can be calculated from the amount that the light is bent.

Some of this missing mass may be accounted for by brown dwarfs. These objects are a link between the planets and the stars, so although they are too small to generate fusion of the hydrogen they contain, they are large enough to generate infra-red energy. This infra-red can be detected by a new range of instruments that effectively brings planets around other stars into view.

The recent infra-red astronomical satellite has enabled astronomers to see much fainter sources of infra-red radiation than ever before. Launched in 1983, it detected infra-red radiation coming from several stars, including Vega, the fifth brightest star in the sky. This radiation is believed to come from large brown dwarf planets, which have surface temperatures of a few thousand degrees. Several of these large planets may be orbiting the stars. There is a growing suspicion among most astronomers that almost every star has something orbiting it.

Since these brown dwarfs can affect the light output from the distant stars, they can be detected by other means. One method is to spot any slight oscillation movement of a star due to the brown dwarf as it obits the star. These efforts have turned up objects with masses equivalent to 10 to 80 times the mass of Jupiter.

At a meeting of the American Astronomical Society in 1989, William Forrest reported the first sighting of a whole group of isolated brown dwarfs. Using an infra-red telescope, he found nine objects bright enough to be brown dwarfs with masses of between five and fifteen times that of Jupiter. These objects were in a nearby region of active star formation, only 450 light years away in the constellation Taurus. Further observations showed that these objects did not orbit any star, but were isolated in space. They were new stars in the making.

Bearing in mind that these observations are just beginning, Forrest believes that brown dwarfs are very common. He says that Taurus may contain ten thousand to a million brown dwarfs, compared with only a few hundred known stars.

Situated towards the outer rim of the Milky Way galaxy is a typical star, our own Sun. The Sun makes its journey at high speed as it revolves around the central bulge of the galaxy, taking about 200 million years for a circuit. As we discussed in the previous chapter, it also bobs up and down along the centre of the galactic disc, accompanied by the planets and their satellites, as well as the various comets, asteroids, meteors and dust which accompany it on its travels. The time for one complete oscillation is about thirty million years, which seems to coincide with mass extinctions.

The Solar System

The scale of our own galaxy is so vast that the Sun and the planets of our own solar system may as well be thought to occupy the same point in space as the Sun. The light from our Sun takes only eight minutes to reach the Earth and then another half a day to reach the orbit of the furthest known planet of our solar system. By contrast the light takes about four years to reach the nearest star, and a further 10,000 years to reach the other side of our galaxy.

Our whole solar system is part of the millions of visible stars which have grouped together in a spinning disc of stars and planets all forming now from the cosmic debris of our local galaxy. Much further out in space there are many other galaxies, all composed of just as many millions of stars, all forming in exactly the same way.

The orderliness of our own solar system is expressed by a simple law discovered in the eighteenth century by Bode. The orbital radius of all the planets follows a definite pattern but outside the order of the planets' gravity lie regions of chaos. Between the planets various cosmic material, ranging from dust grains smaller than a hundred micro-metres to large asteroids hundreds of kilometres across, travels in constantly changing paths for tens to hundreds of thousands of years. The planets, and other dust and asteroids, constantly interact with each other to perturb the orbit of these cosmic objects until an object is eventually captured by one planet. Until recently this theory, known as resonance trapping, would have been difficult to prove, but now the power of modern computers has made it possible to calculate the interactions of hundreds of cosmic bodies and the planets.

It is this process of slow accumulation of matter that has formed our own planet, and the other planets and moons, over eons of time. The result of this slow ongoing expansion is visible on the Earth and on all the other moons and planets of our own solar system.

Chapter 6 - The Solar System

Having observed the differences on a planet due to its size, it should now be possible to sketch in some of the hidden details of the Earth's past. This is the subject of the next chapter.

7 - Ancient Earth

What was the ancient Earth like? Was it boiling hot? Perhaps it was freezing cold? Was there any atmosphere, and if there was, was it a lethal mixture of gases like ammonia or hydrogen sulphide? Why did none of the higher forms of life exist?

All these questions are among the most difficult to answer. In practice, the record of the ancient Earth is so fragmentary that the conditions present during the first 90% of the Earth's lifetime are almost completely unknown. There are only a few intriguing facts.

From the nature of the oldest Earth rocks comes the fact that these rocks formed on an Earth without oxygen. As progressively younger rocks are examined, the oxygen level slowly increases up to its present-day level. The oldest rocks are also less dense than the rocks which tend to form today. They are similar in density to the rocks which are found on the Moon.

The development of life can also be traced in these rocks. At least 3,000 million years ago life had evolved as small single-celled microbes. Then life stayed much the same for eons of time until, about 600 million years ago, suddenly in geological timescales, it exploded into the sea in multi-cellular forms that could swim, see and react.

The evidence available does fit the concept of an Expanding Earth remarkably well. Using the evidence that the Earth has expanded since the time of the dinosaurs it is possible to extrapolate this expansion backwards in time to see how the Earth should have developed since its ancient formation as previously shown in Figure 6.1. An Expanding Earth limits any flights of imagination to within tight boundaries of the possible and impossible, since we can observe the conditions present on

Chapter 7 - Ancient Earth

worlds smaller than ours and use that as a template for the ancient Earth.

Considering the restraint that is imposed by an Expanding Earth, it seems convincing that the known facts about the ancient Earth are explained so well by it. In contrast, a Constant Diameter Earth provides no clear reason why the conditions would change so dramatically.

Atmosphere

This idea of a small ancient Earth helps us to understand the conditions on it, since this ancient Earth was smaller than our present-day Moon, with similar conditions. It was little more than a collection of rock and dust, and because of its low gravitational field it could hold no atmosphere.

This effect is presently occurring on the smallest planets and moons that have little or no atmosphere - as planets increase in size they develop atmospheres. So as the ancient Earth grew in size it began to retain a larger and denser atmosphere. This density has gradually increased until it has reached its present level.

Various evidence shows that the Earth's atmosphere has only recently formed during the latest stage of our planet's development. The time taken for this development is almost unimaginable, since the oldest known Earth rocks have been dated at over 3,800 million years.

Looking back to the very limit of our knowledge of the early Earth, it seems the very ancient Earth had virtually no atmosphere. The evidence about the nature of the very ancient Earth's atmosphere is taken from the rocks which formed then. It is generally thought that the changing composition of these ancient rocks provides a record of the build-up of oxygen in the Earth's atmosphere as shown in Figure 7.1.

In the oldest rocks of the world, compounds have been found which can only occur when free oxygen is not present - when there is no atmosphere. The oldest known examples are small outcrops of sedimentary rocks of pyrite, siderite, magnetite, calcite and dolomite which have chemical compositions which are considered to form only when the oxygen pressure is less than one-millionth of our present atmosphere.

On a Constant Diameter Earth water would easily form into oceans of open bodies of water, but on a Reduced Gravity Earth liquid water could not form until late in its geological history.

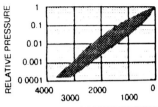

Fig. 7.1 Earth's Developing Atmosphere

The geological evidence from the Earth's ancient crust indicates that the atmosphere has developed over thousands of millions of years.

From its initial creation 4,600 million years ago to about 1,000 million years ago there would have been insufficient gravity to form the dense atmosphere that would stop any large bodies of water boiling away.

Temperature

Today's Earth is held at a reasonably constant temperature by the atmosphere which clings to its surface. Without this layer of air the Earth's temperature would be similar to that of the Moon.

On the Moon the surface temperature rises to 100°C (212°F) in sunlight and falls to -150°C (-238°F) in darkness with an average temperature at the surface of about -18°C (0°F). If the ancient Earth had no atmosphere it should also have similar conditions to the Moon.

Fortunately for present-day life, air traps a large amount of heat to warm our planet by at least 30°C (86°F). In contrast, the temperatures present at the surface of the ancient Earth should have fluctuated widely as the surface was affected by night and day.

Life

To our eyes, the ancient Earth would have looked completely barren. Indeed, it may appear that life could not survive, since there was no atmosphere, no liquid water and no obvious food. But there is evidence to show that the first glimmerings of life are able to survive in such a hostile environment.

One line of evidence for this comes from the present-day meteorites which fall to Earth from time to time. Meteorite types are divided into three broad classes, which depend on the ratio of their metal-to-silicate content - irons, stones and stony-irons.

Chapter 7 - Ancient Earth

One group of stony meteorites which has caused much interest is the carbonaceous chondrites. These meteorites contain up to 5% organic matter and associated hydrated minerals. They are the only known extra-terrestrial material containing organic matter. They are also the oldest known matter and have been dated at 4,500-4,700 million years old.

The carbonaceous material in these meteorites has been held by some to be the result of primitive life within the meteorites. In 1961 Claus and Nagy reported finding small spherical objects in samples of Orgueil and Iruna meteorites.[1] They concluded that these objects may be micro-fossils that lived within the meteorite - they were effectively saying that these meteorites held the first glimmerings of life.

During the 1960s other groups found the same structural bodies in other carbonaceous chondrites. The micro-fossil objects ranged from simple structures to the more complex. Under a scanning electron microscope the organic box-like structures are seen to have five or six sides. The structures are hollow and contain mineral matter which has been interpreted by Fred Hoyle as primitive bacteria or viruses in his book, *The Intelligent Universe.*[2]

Whether or not the carbonaceous chondrites contain life, they certainly contain some of the important chemicals of life. They contain amino-acids, formaldehyde and hydrogen cyanide. They therefore provided large quantities of organic compounds on the ancient Earth's surface.

Other evidence for the ability of life to survive without an atmosphere comes from the study of existing microbes. The reactions which come first in the biochemical sequence appear to have developed early in the history of life and these processes do not use any oxygen - they are anaerobic. As life developed, the oxygen-dependent reactions did not replace the older processes - they were simply added on to the existing anaerobic process. This ancient process can still occur in more modern cells, so when animal muscle cells are deprived of oxygen the cells continue to operate without oxygen at reduced efficiency. The oxygen-using cells have reverted to a more primitive process which does not require oxygen.

From this evidence it is generally concluded that the earliest forms of life are still found amongst the most primitive living cells of today. They are single-celled organisms, *Prokaryotes*,

---

[1] *Claus & Nagy, 1961.*
[2] *Hoyle, 1988.*

Dinosaurs and the Expanding Earth

which lack a nucleus and chromosomes. The bacteria and *Cyanobacteria* (formerly known as blue-green algae) belong to this group.

Bacterial filaments existed on Earth about 3,000 million years ago (and I've already discussed the possibility that these could easily exist on the present-day Mars, and perhaps even the Moon, in Chapter 6). Known as *Stromatolites*, massive cones of micro-fossils have been described from the 2,700 million years old Precambrian rocks in Canada. Well-preserved *Cyanobacteria* filaments forming large *Stromatolite* structures have been found preserved in rocks 2,000 million years old, and this shows that oxygen was being photo-synthetically produced on a vast scale by this time. Considering the early origin of this simple single-celled plant life it has been thought strange that the oxygen atmosphere took so long to develop. If the ancient Earth's gravity was strong enough to retain the oxygen then the Earth would have developed an oxygen atmosphere within a few million years. But we know from the rocks forming at this time that this did not happen. This fits in well with the concept of the oxygen leaking into space due to the reduced gravity on the ancient Earth.

Surprisingly, structures which resemble the *Stromatolites* still exist today as shown in Figure 7.2. They only flourish today in marginal marine environments which are so hostile that most other forms of life cannot exist. Only the organisms which build these Stromatolite-like structures are hardy enough to survive. When the hot baking sun dries them out they become dormant, and when the water returns they come back to life. The best

Fig. 7.2 Stromatolite

Stromatolite-like structures live in areas like Shark Bay, Western Australia, and consist of cones of micro-organisms that form rock-like mounds.

Chapter 7 - Ancient Earth

examples of these Stromatolite-like structures are in tropical areas like the Great Bahama Bank and Shark Bay in Western Australia.

Developing Atmosphere

The various gases which make up the atmosphere of the Earth came from within Earth itself. As the Earth formed, great pressures built up inside it which forced many gases upwards to the surface. This process, which is known as degassing, is still occurring now. The hot springs, geysers and volcanoes of today emit vapour and gases mainly of water, carbon dioxide, sulphur and nitrogen.

The heaviest gases would cling to the ancient Earth first while the lightest gases would tend to escape into space. Today, this effect is still happening as a vast amount of the lightest gas, hydrogen, is generated in the Earth's crust before eventually being lost into outer space.

By simply plotting the relative weights of the different gases now present in the Earth's atmosphere we can gain an impression of which gases would build up first as shown in Figure 7.3. As can be seen, since carbon dioxide is heavier than oxygen the ancient Earth should have had an atmosphere which consisted of mainly carbon dioxide while oxygen leaked into space. As time passed the oxygen level in the atmosphere reached such a level that compounds in the rocks were oxidised as they were formed.

As the oxygen pressure built up within the atmosphere oxidised rocks formed more frequently. Later, other sedimentary rocks began to combine with the higher levels of oxygen in the atmosphere. Notable is the appearance of extensive carbonates and banded iron. Between 3,000 to 1,500 million years ago layers of iron oxides known as Banded Iron Formations, or BIF, formed around the world. Between 1,800 and 1,500 million years ago the deposits of BIF ceased because by this time banded iron began to be oxidised as it was laid down. These deposits began to be laid down as red beds - the Earth literally rusted.

The ancient Earth of about 1,000 million years ago must have looked very similar to the present-day Mars since that is also red due to iron oxide. The atmosphere was also probably similar since it would be mostly carbon dioxide with only small amounts

Fig. 7.3 Gas Relative Weights
Plotting the relative weights of various gases indicates which gases would tend to form on a developing Earth.

of oxygen retained by the reduced gravity as the oxygen slowly escaped into space.

Increasing amounts of carbon dioxide became locked up as carbonates - compounds which include extra oxygen as well as carbon dioxide. This slow accumulation of carbonate rocks through time is explained by the slow formation of the Earth's atmosphere as the ancient Earth's gravity increased. Limestone ($CaCO_3$) and dolomite ($MgCO_3$) are formed from carbon dioxide (CO_2) gas in the Earth's atmosphere and these gases also show how the ancient Earth's atmosphere was becoming more dense. There are few carbonate rocks formed early on the ancient Earth, but the proportion increases later on.

The story of life itself has been related to the build-up of the Earth's atmosphere. The very ancient life forms lived in an oxygen-less environment and were able to break down the rocks to feed themselves. After them developed the *Cyanobacteria*. These single-celled plants produced oxygen as a waste product for thousands of millions of years without the oxygen level of the planet building up. This begins to make sense if the oxygen could escape into space on a Reduced Gravity Earth.

First Cellular Life

It has become apparent that the greatest division between living organisms is between cells with nuclei and those without any nuclei as shown in Figure 7.4. Cells without a nuclei, *Prokaryote*, are simple structures that mostly form simple microscopic life and live in places like soil, hot springs or deep inside rocks.

All the higher forms of life are built up from cells with nuclei and all these cellular animals are different from the principle types of non-nucleated life - bacteria and cyanobacteria.

Judging from the fossil evidence, the modern kind of cell with a nucleus, *Eukaryote*, came into existence about 1,400 million

Chapter 7 - Ancient Earth

Fig. 7.4 Division of Cellular Life

The largest division of life may be between cells without nuclei, Prokaryote (small drawing), and those with nuclei, Eukaryote (large drawing).

years ago. They are the basic building blocks from which all multi-cellular plants and animals are constructed.

Glaciers and the Snowball Earth

Atmospheric pressure has a very marked effect on water. Without the pressure of the column of air above it, no large bodies of liquid water could exist on the Earth. This is because as the pressure is reduced, the water boils. As the atmospheric pressure above a column of water is lowered, small bubbles of steam begin to form and rise to the surface. Soon, the water is in a very turbulent state as it boils at room temperature. At any pressure below 0.6% of the Earth's present atmospheric pressure water can no longer exist as a liquid. It can only exist as solid water-ice or water vapour.

The fundamental properties of water have a profound effect on the development of the Earth. For most of its life liquid water must have been unknown on the Earth since the atmosphere was below a critical pressure. The only water present was frozen deep within the Earth as permafrost but as the atmospheric pressure slowly increased the permafrost crept towards the surface of our Earth until small patches of ice started to form at the coldest places on Earth - the poles. The pressure was still so low that no liquid water could exist. The ice sublimed directly into vapour as the temperature warmed up, and froze back onto the poles as the temperatures fell. The rest of the Earth would be completely devoid of all ice at this time.

These conditions slowly changed over an enormous time - something in the region of 2,000 million years. As the atmospheric pressure built up these ice fields were able to exist at higher temperatures. This enabled the glaciers to creep down

Dinosaurs and the Expanding Earth

and cover virtually the whole Earth from about 1,000 million years ago.

The Snowball Earth is a hypothesis that the ancient Earth's surface became entirely frozen, at least once and perhaps many times before 650 million years ago. Glacial deposits found in tropical regions, such as glacial drop-stones which are typically deposited at the edge of glaciers as they melt, seem to be the most reliable evidence for this.

Alternating layers of these glacial deposits separated by small bands of non-glacial deposits has been interpreted as evidence of glaciers constantly melting and reforming over periods of millions of years. The Snowball Earth seems to be punctuated by warm interglacial periods. As mentioned in the previous chapter during the discussion about Mars, these warm interglacial periods look similar to geological conditions on the present-day Mars.

Most probably this ice-covered planet had isolated pockets of liquid water sealed beneath its surface. Melted water often gathers under today's glaciers to form subglacial lakes. A large number of these subglacial lakes are found under the surface of the Antarctica ice sheet. The largest subglacial lake is Lake Vostok at 250 km (160 miles) long by 50 km (31 miles) wide. The water in the lake appears to have been sealed for at least 15 million years but has remained liquid despite the temperature at the surface dropping to -89°C (-27°F) at its coldest. Photosynthesis can continue under ice up to 100 metres (328 feet) thick and since the water is expected to have high levels of nitrogen and oxygen trapped under the ice it is widely believed that these subglacial lakes will contain life.

Currently the Snowball Earth remains controversial. One problem encountered is how the Earth could become cold enough to totally freeze over and then warm up enough to unfreeze it within a few million years. The Expanding Earth seems to provide an answer to this problem. As explained in Chapter 4, a smaller diameter Earth would result in a smaller temperature difference between the poles and equator and any minor change in global temperature would rapidly affect the whole Earth. A Snowball Earth could rapidly develop and just as rapidly disappear with relatively minor global temperature variations.

This Snowball Earth theory was briefly discussed in relation to Mars in Chapter 6, since Martian conditions should be similar to the ancient Earth. The Snowball Earth implications for the ancient Earth are still being widely discussed.

Several major glaciations are thought to have occurred in the Earth's early history and furrowed surfaces are considered to be the most reliable evidence for this. Three glacial epochs occurred during the late Precambrian, from about 940 million years ago to about 615 million years ago, which were very widespread. Clear indications of these are now found in Greenland, Australia, South Africa, India, China, Canada and the USA. These glaciers of the late Precambrian apparently spread near the equator. For instance, according to rock magnetism data, Australia was located near the equator at the time of its glaciation.

First Seas

At the start of the Cambrian, about 600 million years ago, atmospheric pressure increased sufficiently to allow open bodies of liquid water to form. The glaciers melted and the water flooded inland to form the first seas that the Earth had known. The amount of water released caused extensive flooding of the land areas, especially in the northern hemisphere. The seas slowly crept over the land until they covered over half the present land areas.

The development of the Earth's atmosphere caused the seas to form in two ways. Firstly, the increasing pressure of the atmosphere allowed water to exist in liquid form. An atmospheric pressure just under 0.1 of our present atmospheric pressure for example would allow water to exist until it reached 40°C (104°F) when it would start to boil. Secondly, the increasing density of the atmosphere would begin to trap more heat from the Sun until this increase in temperature melted the Earth's glaciers.

The first person to propose that certain gases trap the Sun's heat was a Frenchman, Joseph Fourier. He was a great mathematician, and he used his own Fourier's theorem to evolve new ideas about the flow of heat. During his studies he discovered the role of carbon dioxide in the atmosphere and he called this the 'hothouse' effect. This is the theory that is now known as the Greenhouse Effect but it has barely altered from the predictions of the turn of the century. Calculations show that without the natural concentrations of carbon dioxide and water in the atmosphere the planet would be 20-35°C (68-95°F) colder than today. Our whole planet would be completely frozen from the poles to the equator without the Greenhouse Effect.

The idea of a greenhouse is simple. As the radiation from the Sun enters the atmosphere it passes through with no difficulty since the gases are completely transparent at visible wavelengths. This radiation is absorbed by the surface of the Earth and then re-emitted as infra-red radiation. Because this radiation has a longer wavelength most of it is trapped by the water and carbon dioxide in the atmosphere. This results in a much warmer lower atmosphere.

Multi-cellular Life

There was a sudden explosion of life in the newly-formed seas 600 million years ago. The rapid increase in life was so great that it has only recently been recognised that life existed before this time as micro-fossils.

Some of the oldest known large fossils in the world are found in Charnwood Forest in the Leicestershire hills of England. About 575 million years ago, primitive soft-bodied animals like *Charnia masoni* began to leave faint impressions of their existence in the Precambrian rocks. They are believed to have been animals similar to present-day corals or jellyfish and were the most highly evolved animals of that time.

As creatures began to form hard shells they began to leave a clear record of their existence. The beginning of the Cambrian period is marked in the fossil record by the appearance of numerous animals with hard outer skeletons, such as aphalopods and silt-loving trilobites, which left an excellent fossil record. The fossil remains of these animals are very similar in lands that are now very far apart. For example, an early Cambrian trilobite, *Redichia*, is found in Australia, eastern Asia and India. Identical cephalopods are found in Australia and western North America, and similar graptolites are found in Wales in Great Britain, and in the USA. Middle Cambrian trilobites were found in North America and the Baltic which are almost identical. The most common explanation for their similarities has been given as migration of these organisms while the continents drifted. But even so, the migration routes are still obscure, unless the Earth is considered to have expanded.

The trilobites suffered at least four mass extinctions before the end of the Cambrian Period. The effects of each of these crises

Chapter 7 - Ancient Earth

also affected the brachiopods - double-shelled animals. The conodonts - ancient eel-shaped swimmers - were also greatly affected by at least one of these mass extinctions.

Our knowledge of the climate improves when the hard-shelled animals spread throughout the oceans of the world. On the Earth of today different species of animals and plants live in different climates. A cold region has a different range of species than a warmer region and this fact has been used to estimate the climate on the ancient Earth.

The Palaeozoic climate (from 570 to 225 million years ago) was generally warm throughout the whole world, from the poles to the equator. I have already explained the reason a smaller diameter Earth would cause this effect in Chapter 4.

First Life on Land

The first suggestion of any land plants appears about 420 million years ago in the fossil rocks from the Silurian period. The land may have been left barren for so long because life could not survive in the thin atmosphere of that time.

The most ancient of plants, the mosses and lichens, can survive extreme environmental conditions. While flowering plants find the conditions too severe on the Himalayan summits 7,000 metres (22,965 feet) above sea level, lichens and mosses live easily. They can survive extreme frost of -24°C (-11°F), and even continue to grow in these extreme conditions. Lichens have survived laboratory temperatures of -200°C (-328°F) by becoming dormant but within minutes they can switch from the dormant state to full activity when the warmth returns. They can live on sterile quartz. They are frequently found in Arctic regions, where the reindeer browse on them during the coldest seasons of the year.

At the other extreme, in the exhausts of volcanoes poisonous gases and boiling acids stream into the atmosphere. On the edge of these infernos have settled lichens and *Cyanobacteria* which can live in hot springs of dilute sulphuric acid at temperatures up to 80°C (176°F).

The assumption that life was stopped from venturing onto land by the thinness of the Earth's atmosphere allows an estimation of atmospheric pressure on the ancient Earth. The highest mountains of the world have an atmospheric pressure of about one third of the atmospheric pressure at sea level and yet life still survives. The atmospheric pressure about 400 million years ago was probably between one tenth to one third the present pressure due to the Reduced Gravity Earth.

Final Thoughts

The most widely held theory about the development of the Earth is that it has maintained the same diameter for the last 4,000 million years. This Constant Diameter Earth is such a strongly held belief that it has become one of the most firmly held scientific dogmas of today.

Scientific dogma often becomes so widely believed that it seems impossible that it could be wrong. History illustrates that many such beliefs are only replaced with great difficulty and we have briefly discussed in Chapters 3 and 4 the long and difficult introduction of both Wegener's Drifting Continents and the Copernican Sun centre system. These new concepts, along with many others, only overturned the scientific dogma of the day by persistent open-minded efforts to highlight the scientific evidence supporting new concepts. Today's dogmatic belief in a Constant Diameter Earth is firmly held despite a wide and diverse range of scientific evidence that the Earth has gradually increased in size, mass and gravity.

This final chapter has covered the first 90% of the history of the Earth's development. The reason for the slow initial evolution of life and atmosphere on the Earth has always been a mystery, but an Earth that was much smaller would be unable to retain an atmosphere to support multi-cellular life during its early geological history.

The concept of a smaller ancient Earth also explains the geological history of the ancient Earth. The geological processes on an ancient Earth can be compared with similar geological processes happening on today's smaller celestial bodies like the Moon and Mars.

The first few chapters explained why the relative scale of life was larger than today on this Reduced Gravity Earth. This is why many dinosaurs were gigantic. Surface gravity can be estimated for the ancient Earth using the relative scale of life, and various other methods, based on scale effects on ancient life.

Chapters 3 and 4 took us through some of the geological evidence for an Expanding Earth. Calculations of ancient gravity using this geological evidence for an Increasing Mass Expanding

Chapter 7 - Ancient Earth

Earth predicted ancient reduced gravity estimates similar to those obtained from the relative scale of life. Both lines of reasoning point to the same conclusion, namely that the Earth is slowly increasing in diameter, mass and surface gravity over geological time.

In Chapter 5 we looked at how the Kant-Laplace Nebular hypothesis for the creation of the Earth could be modified to account for the expansion of the Earth over geological time. The formation of the Earth was not an initial rapid bombardment of cosmic material but the much slower process of an Ongoing Cosmic Accretion from asteroids, comets, meteorites and cosmic dust. This process has lasted for the geological history of the Earth. Today we only see some cosmic showers but over geological time there have been many cosmic storms when the Earth has been bombarded by vast amounts of cosmic material. Most of this cosmic material has been transported into the interior of the Earth by the action of weathering, material transport and subduction.

In Chapter 6 we examined evidence that the same process of mass increase was occurring on other moons and planets. This process can be extrapolated to the formation of whole galaxies to explain why the most ancient red stars are in the centre of galaxies while younger blue stars are still forming in the outer disc of stars.

This diverse range of evidence points to the conclusion that the Earth is still in the midst of creation. The Earth's surface gravity is slowly increasing because the Earth's mass is increasing and this increasing gravity has reduced the scale of life over millions of years. The continents are separating because the Earth is expanding as new mass is added to it. If we look around us with a clear mind, stripped of the dogmas of the past, we can very clearly see these processes happening.

References

Alexander RM, 1976. Estimates of speeds of dinosaurs. Science 261, 129-130.

Alexander RM, 1983. A dynamic similarity hypothesis for the gaits of quadrupedal mammals. J Zool. Lond. 201, 135-152.

Alexander RM, 1989. Dynamics of dinosaurs and other extinct giants. Columbia University Press, New York.

Alden, A. 2010. Why the Expanding Earth animation is really a plate tectonic animation. www.geology.about.com

Alvarez et al, 1980. Extraterrestrial cause for K-T extinction. Science 208, 1095-1108.

Anderson, Rahn & Prange, 1979. Scaling of supportive tissue mass. Q Rev Biol 54:139–148.

Anderson, Hall-Martin & Russell, 1985. Long-bone circumference and weight in mammals, birds, and dinosaurs. J Zool (London) 207:53–61.

Bacon F, 1620. Novum Organum. (Novum Organum is also referred to as Novanum Organum in older translations).

Baker H, 1912. The Origin of Continental Forms.

Bakker R, 1986. The Dinosaur Heresies: New Theories Unlocking the Mystery of the Dinosaurs. William Morrow & Company.

Berends B, 1996. The Expanding Earth. ISBN 1-895305-39-X. Canada.

Borelli G, 1680, 1681. De Motu Animalium (On the Movement of Animals), De Motu Animalium I and De Motu Animalium II.

Bridges, 2002. Our expanding earth, the ultimate cause. ISBN-13: 978-0972409407. Publisher: Oran V. Siler Printing.

Carey SW, 1958. The Tectonic Approach to Continental Drift pp. 177-355 in S. W. Carey (ed): Symposium on Continental Drift, 1956, Univ. of Tasmania, Hobart (Univ. of Tasmania, 1958).

References

Carey SW, 1976. The Expanding Earth. Elsevier Scientific Publishing Company, Amsterdam-Oxford New York.

Carey SW, 1988. Theories of the Earth and Universe. A History of Dogma in the Earth Sciences. Stanford University Press, Stanford, California.

Carey SW, 2000. Earth Universe Cosmos. 2nd Edition. University of Tasmania, Hobart. pp 258.

Chudinov, 1998. Global Education Tectonics of the Expanding Earth. ISBN-13: 978-9067642804. Publisher: VSP International Science Publishers.

Claus & Nagy, 1961. A Microbiological Examination of Some Carbonaceous Chondrites. Nature, 192, 594.

Clube V & Napier B, 1982. The Cosmic Serpent. Universe Pub. ISBN-13: 978-0876633793.

Clube V & Napier B, 1990. The Cosmic Winter. Wiley-Blackwell. ISBN-13: 978-0631169536.

Copernicus N, 1540. On The Revolutions of Heavenly Spheres.

Colbert EH, 1962. The weights of dinosaurs. American Museum Novitates, 2076, pp. 1–16.

Darwin C, 1859. On the Origin of Species. London.

Darwin C, 1890. The Voyage of the Beagle. London.

Darwin C & Wallace A, 1858. On the Tendency of Species to form Varieties; and on the Perpetuation of Varieties and Species by Natural Means of Selection. Linnean Society of London.

Davidson J, 1994. Earth Expansion Requires Increase in Mass. Frontiers of Fundamental Physics. pp295-300.

Davidson J, 2008. Plate Tectonic Structural Geology to Detailed Field and Prospect Stress Prediction. APPEA Journal 2008.

Du Toit, 1937. Our Wandering Continents. South Africa.

Endersbee L, 2000. Reservoirs in Naturally Fractured Rock. ATSE Focus, No. 111 Mar/Apr 2000.

Erickson WC, 2001. On the Origin of Dinosaurs and Mammals.

Farlow JO, 1987. A Guide to Lower Cretaceous Dinosaur Footprints and Tracksites of the Paluxy River Valley, Somervell County, TX. Baylor University, Waco, Texas.

Farlow JO, 1981. Estimates of dinosaur speeds from a new trackway site in Texas. Nature, 294:747-748.

Findlay D, In progress. Plate Tectonics and this Expanding Earth - An Alternative View Incorporating the Earth's spin in geodynamics. E-Book / Disc.

Frank LA, Sigwarth JB & Craven JD, 1986. On the Influx of Small

Comets into the Earth's Upper Atmosphere: I. Observations. II. Interpretation. April Geophysical Research Letters.

Frank LA, Sigwarth JB & Yeates CM, 1990. A Search for Small Solar-System Bodies Near the Earth Using a Ground-Based Telescope: Technique and Observations. Astronomy and Astrophysics, 228:522 (February).

Frank LA, with Huyghe P, 1990. The Big Splash. Published by Birch Lane Press. ISBN 1-55972-033-6.

Galilei G, 1632. Dialogue Concerning the Two Chief World Systems. Roma.

Galilei G, 1638. Discourses and Mathematical Demonstrations Relating to Two New Sciences. Holland.

Gerasimenko and Kasahara, 2002. Tectonic plate motion and deformation by space geodesy measurements (On the question of fixing kinematics reference frame). Geology of the Pacific Ocean, 21 (1), 3-13. (in Russian).

Gillette D. 1991. Seiemosaurus halli, a new sauropod dinosaur from the Morrison Formation (Upper Jurassic/Lower Cretaceous) of New Mexico, USA. Journal of Vertebrate Paleontology 11:417-433.

Halm JKE, 1935. An astronomical aspect of the evolution of the Earth. The Journal of the Astronomical Society of South Africa, Vol. IV (1), 1-28.

Heezen, Tharp & Ewing, 1959. The Floors of the Ocean.

Hess H, 1962, History of Ocean Basins. Geological Society of America. pp 599-620.

Hilgenberg O, 1933. Vom wachsenden Erdball. ('The expanding globe' or 'The expanding Earth'). Munich.

Hilgenberg O, 1962. Palaopollagen der Erde. Neues Jahrb. Geol. und Palaontol., Abhandl 116, Struttgart.

Holmes A, 1944. Principles of Physical Geology. London.

Hoshino M, 1999. The Expanding Earth: Evidence, Causes and Effects. Tokai University Press.

Hoyle F, 1988. The Intelligent Universe. London.

Hoyle & Narlikar, 1971. On the Nature of Mass. Nature, Volume 233, Issue 5314, pp. 41-44.

Hurrell SW, 1994. Dinosaurs and the Expanding Earth. Oneoff Publishing.com. ISBN 0 952 2603 01.

Hurrell SW, 2003. Dinosaurs and the Expanding Earth. Ebook edition. Oneoff Publishing.com. ISBN 0 952 2603 1X.

Hutton J, 1795. Theory of the Earth: with proofs and illustrations.

References

Jeram A, 1990. When scorpions ruled the world. New scientist 16 June 1990. Magazine issue 1721.

Jordan P, 1971. The expanding Earth: some consequences of Dirac's gravitation hypothesis, Oxford: Pergamon Press.

Kuhn T, 1962. The Structure of Scientific Revolutions.

Langston, Lehar, Burke & Weiler, 1990. Galaxy mass deduced from the structure of Einstein ring MG1654+1346. Nature 344, 43 - 45 (01 March 1990).

Lawson DA, 1975, Pterosaur from the Latest Cretaceous of West Texas. Discovery of the Largest Flying Creature. Science, 187, pp. 947-948.

Luckert K, 1999. Planet Earth Expanding and the Eocene Tectonic Event. ISBN 0 967506 0 9. Lufa Studio, Portland, Oregon.

Mann CJ & Kanagy SP, 1990. Angles of repose that exceed modern angles. Geology (Geological Society of America), April 1990. Page 358-361.

Mantovani R, 1889. Reforme du calendrier. Bull. Soc. Sc. Et Arts Reunion. 41-53.

Mantovani R, 1909. L'Antarctide. Je in'instruis. Ka science pour tous, 38. 595-597.

Mardfar R, 2000. The relationship between Earth gravity and Evolution.

Maxlow J, 2005. Terra non firma Earth: plate tectonics is a myth. Terrella Press, Perth, W.A. ISBN 0646449176.

McCarthy, D., 2002. The trans-Pacific zipper effect: disjunct sister taxa and matching geological outlines that link the Pacific margins. Journal of Biogeography. ISSN: 0305-0270. Volume 30: Issue 10. pp1545 - 1561.

McCrea WH,1975. Nature 255, 607-609.

Myers L, 1972. The Accreation of the Earth. St. Clair Enterprises.

Myers L, 1983. The Accreation of the Earth: In Carey, S.W. Expanding Earth Symposium. University of Tasmania.

Myers L, 2006. Open letter to major Science Institutions in America, and various prominent scientists. St. Clair Enterprises. USA.

Myers L, 2010. Accreation - A New Theory of Planetary Creation (How and Why the Earth is Growing and Expanding). St. Clair Enterprises. USA.

Newton I, 1727. The Principia. London.

Novella S, 2009. No growing Earth, but a growing problem with science journalism. Skepticblog.org

Ogrisseg J, 2009. Dogmas may blinker mainstream scientific thinking. The Japan Times Sunday, Nov. 22, 2009.

Owen H, 1979. The paleobathymetry of the Atlantic Ocean from the Jurassic to present: a discussion. J. Geol., 87. 116-118.

Owen H, 1983. Atlas of continental displacement, 200 million years to the present. Published by Cambridge University Press.

Owen H, 1984. The Earth is expanding and we don't know why. New Scientist, Vol. 104, No. 1431, p. 27 - 30.

Parkinson, 1988. (In Carey, 1988 & 2000).

Perin I, 2003. The expanding hemispheric ring. (See Scalera & Jacob, 2003).

Robaudo & Harrison 1993. Plate Tectonics from SLR and VLBI global data. In Smith & Turcotte eds. Contributions of Space Geodesy to Geodynamics: Crustal Dynamics. Geodynamics Series, Volume 23. American Geophysical Union.

Selden P, 1989. Orb-web weaving spiders in the early Cretaceous. Nature 340, 711–713.

Scalera G, 2002. Gravity and Expanding Earth. Proceedings of the 21th National Meeting GNGTS, published on CD-rom, Roma, pp.11.

Scalera G, 2003. The expanding Earth: a sound idea for the new millennium. (See Scalera & Jacob, 2003).

Scalera G, 2004. Gravity and Expanding Earth. In: N.P.Romanovsky (ed.) Regularities of the structure and evolution of Geospheres. Proceedings of the VI interdisciplinar International Sci. Symposium held in Khabarovsk 23-26 Sept. 2003, p. 303-311.

Scalera G, 2006. Review of Dinosaurs and the Expanding Earth. Annals of Geophysics.

Scalera & Jacob, 2003. (Editors). Why expanding Earth?: A book in honour of Ott Christoph Hilgenberg. Publisher INGV, Roma, Italy.

Shapley H, 1949. Galactic Rotation and Cosmic Seasons. Sky and Telescope, volume 9, page 36.

Shehu V, 2005. The Growing and Developing Earth. ISBN-13: 978-1419616631. Publisher: Booksurge.

Sloan, Rigby, Leigh, Van valen & Gabriel, 1986. Gradual Dinosaur Extinction and Simultaneous Ungulate Radiation in the Hell Creek Formation. Science.

Snider-Pellegrini A, 1858. La creation et ses mysteres devoiles. Frank e Dentu, Paris. Pp487.

Taylor FB, 1910. Bearing of the Tertiary Mountain Belt on the Origin of the Earth's Plan. Geological Society of America Bulletin. V21 pp. 179-226.

Thompson D, 1917. On Growth and Form. London.

References

Tschanz K, 1985. Tanystropheus - An Unusual Reptilian Construction, in Konstruktionsprinzipien lebender und ausgestorbener Reptilien (Stuttgart, 1985), Part 4, pp. 169–177.

Vogel K, 1990. The expansion of the Earth - an alternative model to the Plate Tectonics theory. In Critical Aspects of the Plate Tectonics Theory; Volume II, Alternative Theories. Theophrastus Publishers, Athens, Greece, 14-34.

Wegener A, 1915. Die Entstehung der Kontinente und Ozeane. (1st edition only published in German).

Wegener A, 1929. The Origin of Continents and Oceans. (4th edition also published as an English translation).

Wild R, 1980. Tanystropheus (Reptilia, Squamata) and Its Importance for Stratigraphy, Mém. Soc. Géol. France, N. S., No. 139, 201–206.

Wild R, 1987. An Example of Biological Reasons for Extinction: Tanystropheus (Reptilia, Squamata), Mém. Soc. Géol. France, N. S., No. 150, 37–44.

Yarkovsky JO, 1888. Hypothèse cinétique de la gravitation universelle en connexion avec la formation dés éléments chimiques. 134 pp. (Chez l'auteur, Moscou)

Yarkovsky JO, 1899. Universal Gravitation as a Consequence of Formation of Substance Within Celestial Bodies (Russian edition translated). Moscow.

Yeates CM, 1989. Initial Findings from a Telescopic Search for Small Comets Near Earth. Planetary and Space Science, 37:1185 (October).

Index

A

Adams, Neal, 27, 125, 141
Aepyornis titan, 77
Alden, Andrew, 125
Alexander, R. McNeill, 61, 65
Allosaurus, 49, 65
Alvarez, Luis, 151
Alvarez, Walter, 151
Andrewsarchus, 72
Ankylosaurus, 69
Ant, 41
Anteater, 82, 168
Apatosaurus, 29, 65
Ape, 73, 75
Archaeopteryx, 22, 32, 56
Archelon, 70
Armadillo, 69, 74, 82, 168
Asteroid
    Eros, 148
    Ganymede, 148
    Near-Earth, 148
    Potentially Hazardous, 147
    Spaceguard project, 147
Atmosphere
    of ancient Earth, 194
    of Earth, 132, 144, 146, 152, 160, 193, 196, 202, 204
    of giant planets, 186
    of Mars, 181
    of planets, 173
Australopithecus, 75

B

Bacon, Francis, 88
Bakker, Robert, 32, 65
Baluchitherium, 72
Banded Iron Formations, 198
Barnett, Cyril, 85
Bear, 72, 76
Berends, Ben, 110
Blinov, V.F., 85
Borelli, Giovanni, 43
Brachiosaurus, 29, 61
Brontosaurus, 21, See: Apatosaurus
Brosske, Ludwig, 85, 86
Brown dwarf
    becoming a star, 187
    missing mass, 190
    very common, 190
Bruno, Giordano, 114
Bullard, Edward, 102

C

Cambrian
    Mass extinction, 204
    start of, 202
    start of life, 203
    trilobite, 203
Carbonaceous chondrites, 196
Carey, S. Warren, 21, 85, 99, 103, 105, 108, 130, 140, 143
Cave bear, 168
Charnwood Forest, 203
Club moss, 131
    compared to ancient forms, 56
    related to scale trees, 66
Clube, Victor, 143, 164
Cockroach, See: giant cockroach

Index

Coelacanth, 54
Comet, 146, 191
    bombardment, 142, 150
    Encke, 164
    heavy metal layer, 151
Comptosaurus, 66
Conodonts, 204
Constant Diameter Earth, 89, 93, 103, 105, 107, 108, 110, 130, 145, 158, 177, 194, 205
Constant Mass Expanding Earth, See: Expanding Earth
Continental
    blocks, 86
    drift, 86, 105
    erosion, 87
    refitting on smaller Earth, 105
    single supercontinent, 89
Convection Cell, 97, 103, 104, 107, 159
Cope, Edward, 31
Copernican system, 114
Copernicus, Nicolaus, 113
Cosmic dust, 142, 146, 171
    dense area of, 154
    disc of, 153
    heavy metal layer, 151
Cox, Allan, 85
Creer, Kenneth, 85
Cretaceous, 69
    deposits in spain, 55
    extinction at end of, 70
    North America, 69
Cyanobacteria, 197, 199, 204
Cypress, 67

D

Dachille, F., 143
Darwin, Charles, 73, 81
Davidson, John K., 21, 23, 109
Deccan Traps, 160
Deer, 40, 49
Deinonychus, 31
Dicynodon, 106
Dinohyus, 72
Diplodocus, 21, 29, 65
Dirac, Paul, 100
Doell, Richard, 85

Dragonfly, See: Giant dragonfly, 46, 48
du Toit, Alexander, 95

E

Egyed, Lazio, 85
Einstein Ring, 189
Elephant, 29, 36, 40
Endersbee, Lance, 25
Erickson, William Carnell, 25
Eriksson, Leif, 169
Ewing, Maurice, 98
Expanding Earth
    Carey, 21, 85, 105
    Constant Mass, 22, 24, 139
    Continental Drift, 101
    Davidson, 23
    gravity, 23
    Halm, 97
    Heezen, 103
    Hilgenberg, 85
    Increasing Mass, 22, 24, 27, 84, 115, 116, 127, 129, 139, 146, 159, 167, 172, 206
    Jordan, 100, 175
    Mars, 179
    Maxlow, 24
    Moon, 176
    Owen, 105, 141
    proposed by, 84
    reconstructions, 124
    rediscovery, 85
    subduction, 159
    uniform temperatures, 133
    Yarkovsky, 85
Expanding Universe, 188
Extinction, See: Mass extinction

F

Forrest, William, 190
Fossil spider, 55
Fourier, Joseph, 202

G

Galilei, Galileo, 42, 113
Geology Today, 21
Giant

Armadillo, 74
Beaver, 72, 74
Cockroach, 56
Dragonfly, 48, 58
Irish elk, 72
Kangaroo, 74
Mayfly, 56
Millipede, 48, 57, 58, 131
Platypus, 74
Sloth, See: Megatherium, 74
South America animals, 73
Warthog, 73
Gillette, David, 62
Glossopteris, 93
Glyptodon, 69, 74
Graptolites, 203
Greenhouse Effect, 202
Groves, Ralph, 85

H

Halm, J., 85, 97
Heezen, Bruce, 85, 98, 103
Hellas Planitia, 182
Hess, Harry, 103
Hilgenberg, Ott Christoph, 26, 86, 94, 95, 125, 130, 141
Holmes, Arthur, 97, 103
Homo erectus, 75
Horsehead Nebula, 152, 188
Horsetail plant, 48
Hoyle, Fred, 101, 140
Hubble, Edwin, 187
Humason, Milton, 187
Hummingbird, 46
Hutton, James, 87
Huxley, Thomas, 31

I

Ice Age, 78, 188
Iguanodon, 22, 29, 66
Increasing Mass Expanding Earth, See: Expanding Earth
Indricotherium, 72
Insect, 44, 45, 49
Irish elk, See: Giant Irish elk
Isolated Subduction, 158, 159, 160

J

Jensen, Jim, 62
Jeram, Andrew, 55
Jordan, Pascual, 85, 100, 175
Jupiter, 186
Jurassic, 69
  arid conditions, 135
  Atlantic ocean formation, 134
  break-up of continents, 135
  largest known dinosaur, 61
  sauropod dinosaurs, 65
  sea level, 134

K

Kanagy, Sherman P., 22
Kangaroo, See: Giant Kangaroo
Kant-Laplace Nebular hypothesis, Ongoing Cosmic Accretion, 144, 146, 206
Keindl, Josef, 85
Kepler, Johannes, 114
King, Lester, 85
Kirillow, 85, 86
Kraus, Werner, 58
Kuhn, Thomas, 111

L

Large birds, 77
Lion, 49
Luckert, Karl W., 85, 125, 130

M

Mammoth, 73, 168
Man, C. John, 130
Mann, C. John, 22
Mantovani, Roberto, 85, 88
Mardfar, Ramin Amir, 25
Mars, 179
  Life, 183
  Methane plumes, 185
Marsh, Charles, 31
Mass extinction, 35
Matthews, Drummond, 102
Mauna Loa, 182
Maxlow, James, 24, 108, 109, 124, 130

Index

Mayfly, See: Giant Mayfly
McCarthy, Dennis, 111
McCrea, W.H., 162
Meganeuropsis permiana, 58
Megatherium, 74
Mesosaurus, 93
Mesozoic Era, 61
    gigantic life, 35
    last 50 million years, 68
    plants, 66
    Pterosaurs, 67
Meteorite
    types of, 195
Mid-ocean ridge, 99, 102, 104, 136, 160, 173
Milky Way Galaxy, 189
Millipede, See: Giant Millipede, 58
    astonishing size of, 48
    similar to ancient forms, 80
Museum
    Berlin, 29, 62
    British, 105
    Clausthal-Zellerfeld, 58
    Glasgow, 57
    New Mexico, 56
Myers, Lawrence S., 143

N

Napier, Bill, 143, 164
Neptune, 186
Newton, Isaac, 127
Noel, David, 111
Novella, Steven, 107

O

Ogrisseg, Jeff, 107
Olympus Mons, 182
Ongoing Cosmic Accretion, 146, 178, 206
Optimum form, 41, 43, 45, 47
Ostrom, John, 31
Owen, Hugh G., 20, 105, 125, 130, 141
Owen, Richard, 31

P

Perin, Ilton, 110, 130

Perry, Ken, 125
Phororhacos, 72
Platypus, See: Giant platypus
Pleistocene, 72
Precambrian
    faults in, 96
    Ice Ages, 202
    similar to Mars conditions, 180
Pteranodon, 67
Pterosaur, 61, 67

Q

Quetzalcoatlus, 21, 67

R

Reduced Gravity Earth, 21
    Adams, 27
    ancient atmosphere, 199, 205
    ancient sediment evidence, 22
    Carey, 22
    change in gravity estimate, 81
    controversy, 20
    dinosaurs' gigantic size, 19
    Endersbee, 26
    Expanding Earth, 116, 129, 139
    Giancarlo, 26
    interest in, 27
    Internet discussion, 27
    larger scale of life, 80
    liquid water formation, 194
    relative scale of life, 205
Rhinoceros, 40, 49

S

Sabre-tooth cat, 73, 168
Salamander, 60
Saturn, 186
Sauropods
    declining, 68
    in water, 34
    largest known, 61
    long necks, 32, 49
    most gigantic, 29
    neck forces, 65
Scalera, Giancarlo, 26, 130
Seismosaurus, 62
Selden, Paul, 55

Shapley, H., 162
Siberian, 204
Size and gravity relationship, 51
Slipher, Vesto Melvin, 187
Sloan, Robert, 70
Sloth, See: Giant Sloth, 82, 168
Snider-Pellegrini, Antonio, 88
Snowball Earth, 184, 201
Snowball Mars, 184
Solar System, 191
Spider, 57
Stalactites and stalagmites, 157
Stegosaurus, 66
Stromatolites, 197
Subduction zone, 104, 105, 106, 110, See: Isolated Subduction, 142, 172, 206
Sun
   cooling of Earth, 161
   effect of cosmic dust, 171
   greenhouse effect, 132, 203
   higher constant of gravity, 101
   orbit of, 113
   rotation around galaxy, 154, 189
   scale of, 191
   solar output increase, 180
Supersaurus, 62

T

Tanystropheus, 33, 50
Taylor, Frank B., 89
Terrible lizard, 29
Tethys Sea, 105
Tharp, Marie, 98
Thompson, D'Arcy, 48
Tiger, 49
Titanothere, 72
Transform fault, 117
Triassic
   gravity, 24
   plants, 66
   reptiles, 106
   sea level, 134
   sediments, 33
Triceratops, 21, 68
Trilobite
   mass extinction, 204
   Redichia, 203

Tunguska impact, 164
Tyrannosaurus rex, 21, 32, 69
   bone cells, 56

U

Uintatherium, 72
Ultrasaurus, 62
Universal Constant of Gravity, 84, 100, 101, 127
   definition of, 83
Uranus, 186

V

Valles Marineris, 179
Van Hilten, D, 85
Venus, 185
Vertebrae, 33
Vine, Fred, 102
Vogel, Klaus, 85, 125, 130

W

Warthog, See: Giant Warthog
Water-lily, 42
Wegener, Alfred, 90
Wildebeest, 49

Y

Yarkovsky, Jean, 85, 88